Springer Theses

Recognizing Outstanding Ph.D. Research

Aims and Scope

The series "Springer Theses" brings together a selection of the very best Ph.D. theses from around the world and across the physical sciences. Nominated and endorsed by two recognized specialists, each published volume has been selected for its scientific excellence and the high impact of its contents for the pertinent field of research. For greater accessibility to non-specialists, the published versions include an extended introduction, as well as a foreword by the student's supervisor explaining the special relevance of the work for the field. As a whole, the series will provide a valuable resource both for newcomers to the research fields described, and for other scientists seeking detailed background information on special questions. Finally, it provides an accredited documentation of the valuable contributions made by today's younger generation of scientists.

Theses are accepted into the series by invited nomination only and must fulfill all of the following criteria

- They must be written in good English.
- The topic should fall within the confines of Chemistry, Physics, Earth Sciences, Engineering and related interdisciplinary fields such as Materials, Nanoscience, Chemical Engineering, Complex Systems and Biophysics.
- The work reported in the thesis must represent a significant scientific advance.
- If the thesis includes previously published material, permission to reproduce this must be gained from the respective copyright holder.
- They must have been examined and passed during the 12 months prior to nomination.
- Each thesis should include a foreword by the supervisor outlining the significance of its content.
- The theses should have a clearly defined structure including an introduction accessible to scientists not expert in that particular field.

More information about this series at http://www.springer.com/series/8790

Stefano Salvatore

Optical Metamaterials by Block Copolymer Self-Assembly

Doctoral Thesis accepted by the University of Cambridge, UK

Author
Dr. Stefano Salvatore
Cavendish Laboratory
Department of Physics
University of Cambridge
Cambridge
UK

Supervisor
Prof. Ullrich Steiner
Cavendish Laboratory
Department of Physics
University of Cambridge
Cambridge
UK

ISSN 2190-5053 ISSN 2190-5061 (electronic)
ISBN 978-3-319-38278-4 ISBN 978-3-319-05332-5 (eBook)
DOI 10.1007/978-3-319-05332-5

Springer Cham Heidelberg New York Dordrecht London

© Springer International Publishing Switzerland 2015
Softcover reprint of the hardcover 1st edition 2015
This work is subject to copyright. All rights are reserved by the Publisher, whether the whole or part of the material is concerned, specifically the rights of translation, reprinting, reuse of illustrations, recitation, broadcasting, reproduction on microfilms or in any other physical way, and transmission or information storage and retrieval, electronic adaptation, computer software, or by similar or dissimilar methodology now known or hereafter developed. Exempted from this legal reservation are brief excerpts in connection with reviews or scholarly analysis or material supplied specifically for the purpose of being entered and executed on a computer system, for exclusive use by the purchaser of the work. Duplication of this publication or parts thereof is permitted only under the provisions of the Copyright Law of the Publisher's location, in its current version, and permission for use must always be obtained from Springer. Permissions for use may be obtained through RightsLink at the Copyright Clearance Center. Violations are liable to prosecution under the respective Copyright Law.
The use of general descriptive names, registered names, trademarks, service marks, etc. in this publication does not imply, even in the absence of a specific statement, that such names are exempt from the relevant protective laws and regulations and therefore free for general use.
While the advice and information in this book are believed to be true and accurate at the date of publication, neither the authors nor the editors nor the publisher can accept any legal responsibility for any errors or omissions that may be made. The publisher makes no warranty, express or implied, with respect to the material contained herein.

Printed on acid-free paper

Springer is part of Springer Science+Business Media (www.springer.com)

Supervisor's Foreword

Metamaterials are among the most exciting recent developments at the interface of materials science and physics. The periodic assembly of structural elements of conventional materials on micrometer and sub-micrometer length scales result in unusual wave propagation, endowing the material with properties that cannot be found in nature. Wave propagation within the materials is no longer determined by the intrinsic properties of the structural elements but reflects the symmetry and geometry of the assembly. While this applies to the propagation of any wave, much attention is paid to the propagation of light. The most significant property of metamaterials is the reversal of their refractive index, enabling the creation of superlenses that are able to focus light beyond the diffraction limit, i.e. beyond what is possible with conventional lenses.

The structural requirements of metamaterials are difficult to meet. They necessitate detailed control over 3D materials assembly with a resolution smaller than the wavelength of interest. This is particularly challenging for *optical* metamaterials that mould the flow of visible light. Here, the structural motifs are on the order of 10–50 nm. Since "top-down" methodologies for the assembly of materials with these structural feature sizes do not exist, novel strategies are required.

Stefano Salvatore's doctoral thesis makes a substantial contribution to the field of optical metamaterials. It harnesses the self-assembly of block-copolymers for the creation of 3D morphologies that are promising for metamaterials and explores the reliable replication of the self-assembled morphology into 3D periodic networks of gold struts that support the propagation of plasmon polaritons. The obtained results are spectacular: generating a network of extremely fine struts, light transmission of more than 30 % was achieved through layers that are several hundred nanometers thick. Light transmission through a thick gold layer is the hallmark of the metamaterial effect; since gold layers thicker than a few nanometers have zero transmission, transparency of thicker layers is evidence for the existence of electronic modes (plasmons) transporting the optical excitations across the film. An important aspect of this network morphology is its structural chirality, which is predicted to lie at the core of a possible negative refractive index for visible light.

The thesis is a very careful study, enhancing our knowledge of how optical metamaterials work. It pioneers a manufacturing method that improves the quality of the nanostructured layers, producing optical signals of unprecedented clarity. It goes on to scan the parameter space that determines the optical signature of the material, i.e. the lattice periodicity, strut width, lattice symmetry, strut material and dielectric material surrounding the struts. A further refinement of the lattice structure was obtained by the development of a method that creates hollow struts. These "waveguide cables" reduce the losses in plasmon-polariton propagation, resulting in a stunning 60 % transmission across a 300 nm thick layer. The final chapter of the thesis demonstrates the use of gold metamaterials as sensitive vapour sensors.

Overall, this thesis not only demonstrates a viable route for the manufacture of 3D optical metamaterials at visible wavelengths, it also provides new insights into their functioning. Looking ahead, this thesis provides a sound basis for the further development of self-assembled nanomorphologies as a step towards the creation of materials that have a negative refractive index at visible wavelengths.

Cambridge, May 2014 Prof. Ullrich Steiner

Abstract

This thesis explores the fabrication processes and the optical characterization of metamaterials made by block copolymer self-assembly.

Optical metamaterials are artificial systems designed to produce optical properties that may not be found in nature. They gain their properties not from their chemical composition but rather from their design structure that affects the interaction with electromagnetic waves, producing an optical response different from that of the constituent material. The structural features are of sub-wavelength size so that the metamaterial is perceived homogeneous by the incident waves and the electromagnetic response is expressed in terms of homogenized material parameters.

In this work such structural features will be fabricated well below optical wavelengths by metal replica of a gyroid morphology generated by block copolymer self-assembly.

Block copolymers consist of two or more chemically different polymers that are covalently tethered. Self-assembly in such systems is driven by enthalpy reduction through microphase separation that minimizes unfavourable interfaces, leading to a range of potential morphologies. The gyroid morphology consists of a continuous three-dimensional structure with constant mean curvature and chiral directions.

The gold gyroid effectively behaves as a metamaterial with its own distinct optical characteristics: a reduced plasma frequency and highly enhanced transmission, a hallmark of optical metamaterials. The optical characterization discussed in this dissertation showed good agreement with finite difference time domain (FDTD) calculations and analytical models.

The optical properties of the fabricated metamaterials were successfully tuned by modifying the structural dimensions and the surrounding mediums. The transmission efficiency was then further enhanced by fabricating a hollow gyroid structure with increased surface area. The hollow gyroid enabled also the fabrication of composite metamaterial employing the combination of different metals in the 3D continuous structure.

Next, flexible and stretchable metamaterial were fabricated by infiltrating an elastomer around the gyroid network, paving the way for practical applications.

Finally, the gyroid metamaterial was used as vapour sensor, exploiting the regular pore size and the variation of the optical response with surrounding media.

Contents

Chapter 1
Introduction

> *Either a body absorbs light or it reflects or refracts it or does all these things. If it neither reflects nor refracts nor absorbs light, it can not itself be visible.*
>
> (The Invisible Man, George H. Wells, 1897)

In G. Wells' famous novel, a scientist invented a way to change the refractive index of his body to that of air so that it adsorbed and reflected no light, becoming invisible. Besides its science fiction, it is interesting to consider how the optical properties of matter can be controlled and engineered. The optical properties of any material result from the electric and magnetic responses of its constituent atoms to the electromagnetic waves of light. Such response can be tailored only to some extend by adjusting the chemical composition.

However, recently it has been realized that the internal microstructure of a material can be just as important as the chemistry in determining its optical properties. In fact, by exploiting both chemistry and microstructure, materials can be produced with properties never found in nature, and can, in principle, alter light propagation leading to a cloaking device. This new class of materials, called metamaterials, achieve this effect through a structure that is much smaller than the wavelength of radiation in the region of interest, so that the electromagnetic radiation interacts with the metamaterial as if it was a continuous material.

In this way, the overall properties are driven by the electric and magnetic field induced by the engineered structure, producing a response that may be different from the constituent material.

Several structures have been proposed in the last few years and a wide range of outstanding properties have been successfully produced in the microwave region. Nonetheless, a metamaterial active at optical wavelengths (from about 400 to 700 nm) is radically more difficult to produce, due to the required scale of the structural features.

In this work a gold metamaterial is fabricated with feature sizes below 10 nm and, remarkably, it is created by self-assembly. The fascinating architecture of this metamaterial, called gyroid, is obtained by block copolymer self-assembly.

© Springer International Publishing Switzerland 2015

S. Salvatore, *Optical Metamaterials by Block Copolymer Self-Assembly*, Springer Theses,
DOI 10.1007/978-3-319-05332-5_1

The first chapter of this dissertation will focus on the theoretical background of the block copolymer self-assembly and on the metamaterial theory. I will also dedicate a large section to the gyroid architecture and to its peculiar geometrical properties.

I will begin the discussion of my experimental results in Chap. 2 describing the fabrication process. This Chapter highlights the largest amount of work that went into the fabrication of the gyroid metamaterials and to the development of a solid and reproducible fabrication process.

The optical characterization will then be discussed in Chap. 3. The gyroid metamaterial has raised in the past few years increasing attention in the metamaterial research community. Although there are no other research groups experimentally producing gyroid metamaterials, there have been several theoretical studies predicting interesting properties of this structure. These theoretical results will often be mentioned through out this dissertation to support my experimental results and help the understanding of the gyroid optical properties.

As already mentioned, the properties of optical metamaterials are strictly connected to their structure. In Chap. 4, I will show the tuning mechanisms to tailor the optical properties of the gyroid metamaterial by altering the structural parameters and the surrounding medium.

A further development in the gyroid fabrication has led to the so-called hollow gyroid, presented in Chap. 5. The gyroid morphology was produced with an additional internal interface, giving rise to an improvement of the optical properties and to an extra degree of freedom in designing an optical metamaterial.

In Chap. 6, I will show the fabrication and the optical properties of a flexible and stretchable metamaterial produced by infiltrating an elastomer into the gold gyroid network. Although the optical properties varied only marginally, the importance to produce a metamaterial not confined to flat films is significantly relevant for any practical use.

An application of the gyroid metamaterial will then presented in Chap. 7. The gold gyroid was used as vapour sensor, exploiting the high monodispersity of the pore size and the strong optical variation with respect to different surrounding media.

Finally, the conclusions of this work will be summarized and an outlook will be given in the final chapter with the hope to get closer to the vision of George Wells and his novel.

Chapter 2
Background

This work is based on the combination of two distinct subjects: the self-assembly of block copolymers in thin films and optical metamaterials.

In this chapter I will give a brief introduction of these two fields and of their recent research progress.

2.1 Block Copolymers

Polymers are a class of macromolecules composed of many repeating subunits, defined as monomers, linked by covalent bonds. Polymers can be classified into two types: homopolymers composed of only one repeating monomer unit, and copolymers derived from two or more types of monomer units. In turn, copolymers can be classified based on how these units are arranged along the chain. The arrangement can generate: alternating copolymers with regular alternating A and B units (e.g. A-B-A-B-$\cdots$), periodic copolymers with A and B units arranged in a repeating sequence (e.g. (A-B-B-A-B)n), statistical copolymers in which the sequence of monomer residues follows a statistical rule, and block copolymers [1]. In the block copolymer subclass, two or more homopolymers are covalently arranged into distinct blocks [2]. The simplest example is an A-B block copolymer with two covalently tethered homopolymers. Block copolymers with two or three distinct blocks are called diblock copolymers and triblock polymers, respectively (Fig. 2.1).

Block copolymer can be synthesized by living polymerization [3], a technique that enables the sequential addition of monomers and the synthesis of functional-ended polymers by selective termination allowing the additive synthesis of additional blocks.

An important property of block copolymers is their ability to self-assemble into periodic nanostructures. To understand the block copolymer self-assembly let us first consider a mixture of two separate immiscible homopolymers A and B. The minimization of unfavourable interfaces drives the homopolymers to blend to macrophase separation into coexisting bulk phases, similarly to water and oil mixtures. When two

© Springer International Publishing Switzerland 2015
S. Salvatore, *Optical Metamaterials by Block Copolymer Self-Assembly*, Springer Theses,
DOI 10.1007/978-3-319-05332-5_2

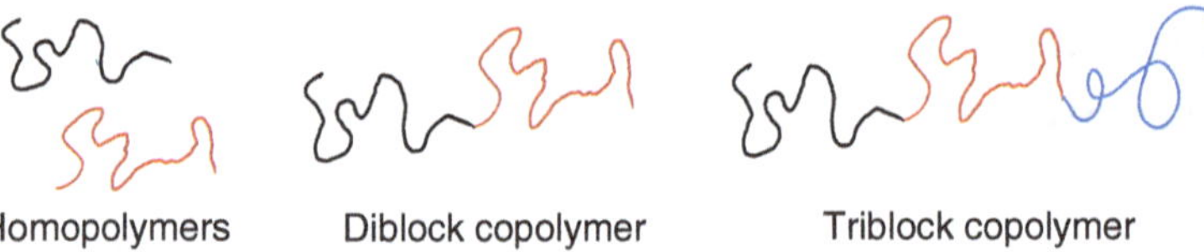

Fig. 2.1 Homopolymers and block copolymers architectures. Polymers are a class of macromolecules that are composed of many repeating subunits. Homopolymers consist of a single type of monomer, while linear block copolymers are chains containing sequences of different homopolymers

immiscible homopolymers are covalently bonded to form an A-B block copolymer, the length scale of the phase separation is limited by the polymer chain length, leading to microphase separation.

In the block copolymer melt, the enthalpic forces, driving the interface minimization, are counterbalanced be entropic forces that encourage the reduction of chain stretching [4]. The affinity between two polymers is defined by the Flory-Huggins interaction parameter χ. In addition to the Flory-Huggins parameter, the phase behaviour of block copolymers is controlled by the degree of polymerization N and the volume fraction of each block (f_a and f_b in a diblock copolymer).

For volume fractions $f_\mathrm{a} \simeq f_\mathrm{b}$, the interface minimization drives the phase separation to lamellae containing alternating the A and B blocks. For volume fractions $f_\mathrm{a} \gg f_\mathrm{b}$ the best geometrical solution to minimize the polymer interface leads to the formation of spheres containing the polymer with the lower volume fraction, embedded in a major phase of the other polymer.

The equilibrium phases can therefore be mapped as a function of the volume fraction f_a and the χN parameter in a phase diagram, shown in Fig. 2.2.

2.1.1 Triblock Copolymer Morphologies

The addition of a third block to a linear block copolymer gives rise to several new potential architectures, some of which are shown in Fig. 2.3, with many others still under investigation [6–8].

In the gyroid morphology, the blocks forming the two interwoven networks are composed of distinct polymers, denoted as minority phases, each forming a so-called single gyroid. The largest part of the work presented in this dissertation focuses on gyroid formed by triblock copolymers, where one of the minority phases was selectively etched and backfilled by electrodeposition, producing a metal single gyroid.

The single gyroid has strong chirality in certain directions, with helices visible in the [111] and [100] orientations as shown in Fig. 2.4. In these directions we can distinguish two helices with opposite chirality and different radii.

The phase map of a triblock copolymer composed of polyisoprene (PI), polystyrene (PS) and polyethylene oxide (PEO), commonly named ISO block copolymer, is

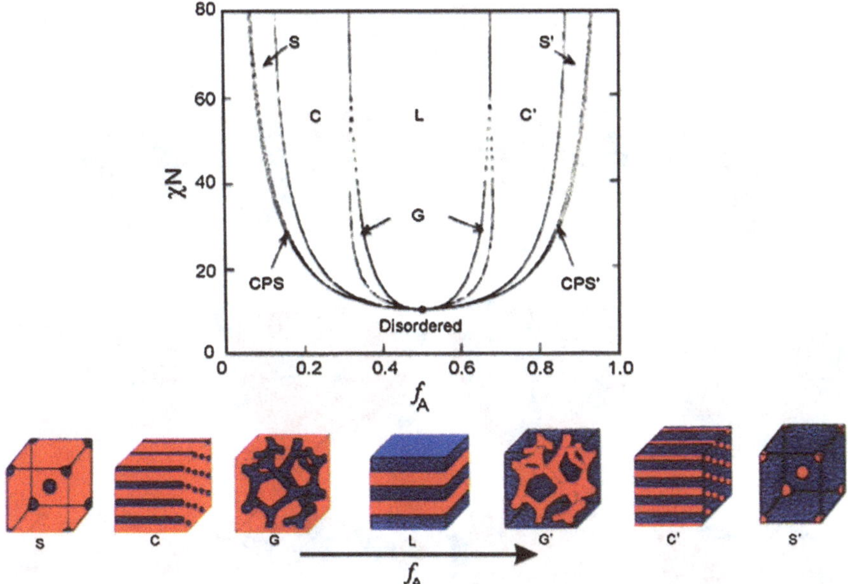

Fig. 2.2 Phase diagram for linear AB diblock copolymers with the regions of stability for *cylinders* (*C*), *spheres* (*S*), *lamellae* (*L*), *gyroid* (*G*), *close packed spheres* (*CPS*), and disordered phase. Adapted from [5]

displayed in Fig. 2.5. As it will be discussed in the next chapter, the metal single gyroid will be fabricated from the ISO triblock copolymer with the alternating gyroid morphology. This morphology is generated in a narrow area of the phase diagram as displayed in Fig. 2.5.

2.1.2 Gyroid Morphology

The gyroid phase observed in diblock copolymers is a structure consisting of two interwoven continuous networks, containing the polymer with lower volume fraction, embedded in a matrix formed by the majority block. Since both minority and majority phases are continuous in all three dimensions the structure is termed bicontinuous.

The term "gyroid" was coined in 1970 by Alan Schoen [12], a mathematician who discovered the family of triply periodic minimal surfaces (i.e. with zero mean curvature everywhere), which are now named after him. The "Schoen G" or "gyroid surface" is a triply periodic minimal surface that divides the space into two separate twisting volumes as shown in Fig. 2.6a.

The majority phase of the gyroid block copolymer is equivalent to the gyroid surface with a given finite thickness, by simultaneously expanding of the two surfaces on both sides. Space is consequently divided into three non-intersecting volumes. The

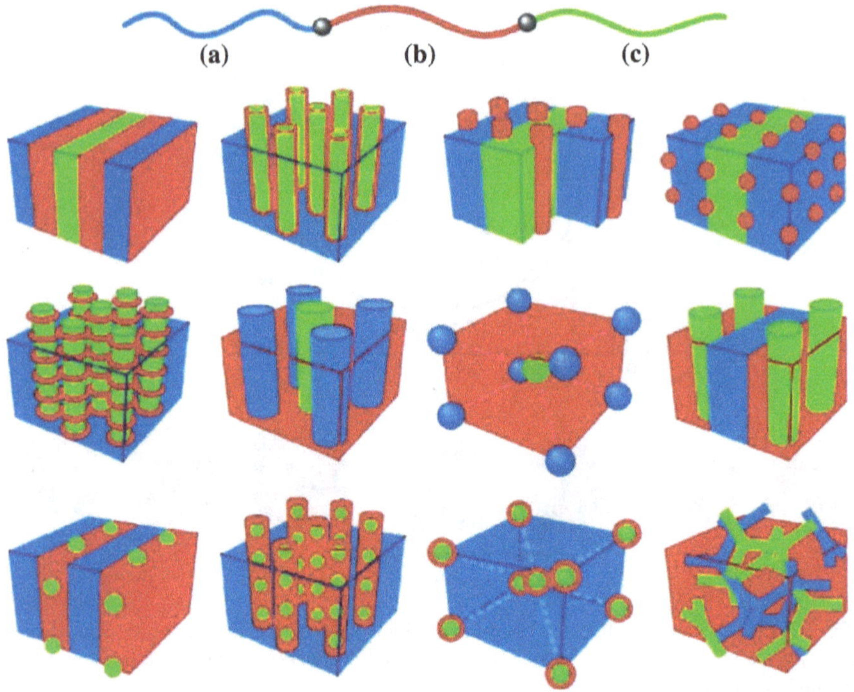

Fig. 2.3 Morphologies for linear ABC triblock copolymers. The phases are coloured as shown by the copolymer blocks at the *top*, with block types A, B and C confined to regions colored *blue*, *red* and *green*, respectively. Adapted from [9]

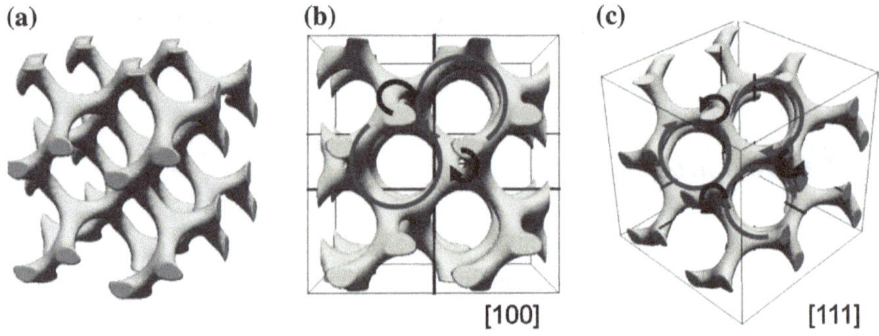

Fig. 2.4 **a** Perspective view of the gyroid structure with 10 % filling fraction. **b** View of the chiral [100] direction with the two opposite helices indicated by the *black* and *red arrows*. **c** View of the chiral [111] direction. Adapted from [10]

"central volume" forms the majority phase, embedding two distinct non-intersecting gyroid networks. The two gyroid network morphology will be hereafter called double gyroid.

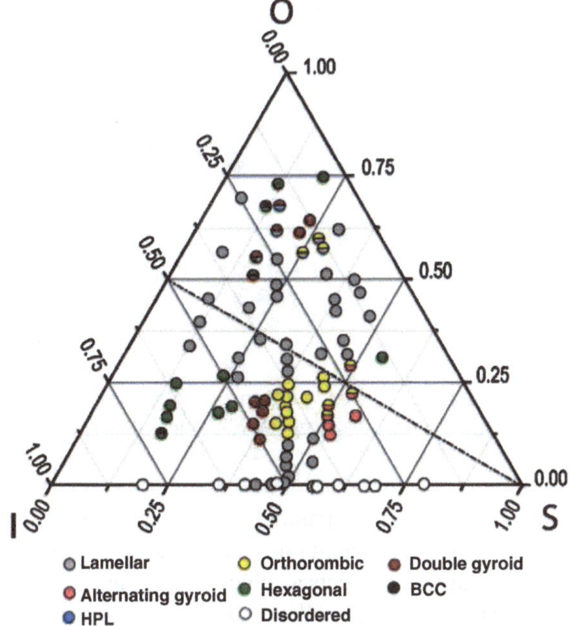

Fig. 2.5 Experimental phase map of the triblock copolymer systems composed of polyisoprene (*I*), polystyrene (*S*) and polyethylene oxide (*O*) at a set value of χN [11]. The axes identify volume fractions of each block. The phase morphologies are identified by colours and labeled below the diagram with abbreviations defined as: LAM (lamellar), HEX (hexagonal), BCC (body-centered cubic), HPL (hexagonally perforated layers). The schematic representations of these morphologies are described in [11]

Why polymers select this non trivial morphology is an interesting issue that can be addressed by examining the concept of minimal surface. A minimal surface is a surface that minimizes the total surface area under some constraint. For example,

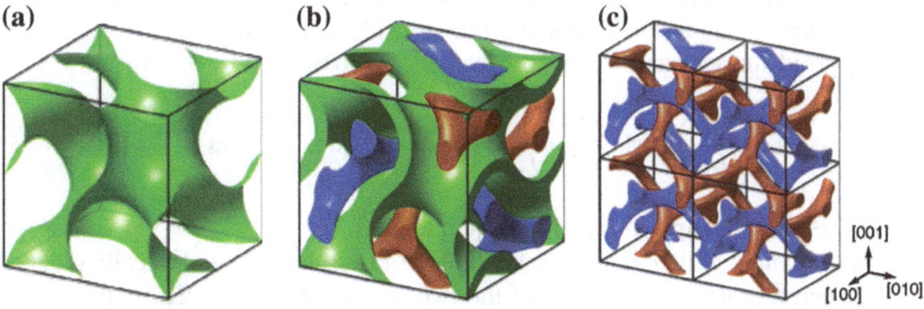

Fig. 2.6 **a** "Schoen G" or "gyroid surface". **b** A schematic of the the double gyroid unit cell showing the majority phase equivalent of the gyroid surface (*green*) and the two gyroid networks (*blue* and *red*). **c** Four cubic unit cells of the double gyroid network with a volume fraction of 12 %. Adapted from [13]

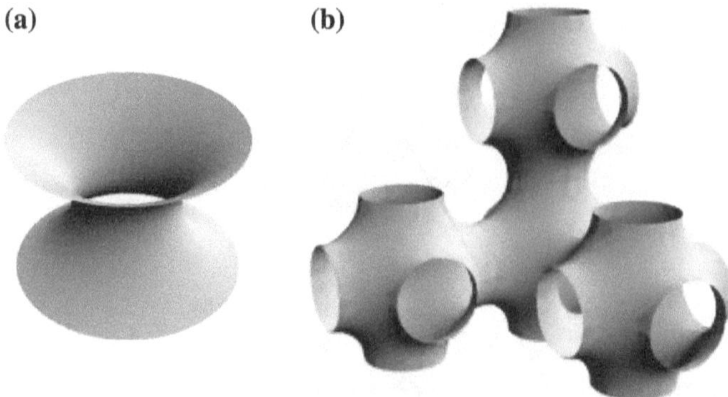

Fig. 2.7 **a** Catenoid minimal surface derived by minimizing the connecting surface of two parallel rings. **b** Schwartz P periodic minimal surface produced by connecting a cubic array of rings

under the constraint of connecting two parallel rings, the geometry that minimizes the surface area corresponds to the catenoid minimal surface. Similarly, the Schwartz P triply periodic minimal surface connects a cubic array of rings, as shown in Fig. 2.7. The gyroid surface can be seen as a tessellation of gyrating catenoids associated with the P-surface [14].

In the gyroid block copolymer system, the unfavourable interaction between polymers leads to the segregation of majority block in the the central matrix volume that, as a minimal surface, represents an optimized geometry which minimizes interface interactions.

Interestingly, the gyroid geometry is also found in nature in several systems in addition to block copolymers. In 1983 it was found in some surfactants, where lipid appeared to form a single bilayer continuous in three dimensions, separating two networks of water with a Schwartz G geometry [15].

Gyroids have been observed in biological structural coloration in butterfly wing scales [16, 17]. In these systems the gyroid structure has lattice size one order of magnitude higher than in the block copolymers, giving rise to photonic crystals. Gyroid structures are also occasionally found inside cells [18].

The gyroid surface can be trigonometrically approximated by the equation:

$$\sin\left(\frac{2\pi}{L}x\right)\cos\left(\frac{2\pi}{L}y\right)+\sin\left(\frac{2\pi}{L}y\right)\cos\left(\frac{2pi}{L}z\right)+\sin\left(\frac{2\pi}{L}z\right)\cos\left(\frac{2\pi}{L}x\right)=t, \quad (2.1)$$

where L is the cubic unit cell and t is equal to 0.

For increasing t with $0 < |t| \leq 1.413$ the gyroid surface is monotonically reduced with an increase of volume of one of the two space regions divided by the gyroid surface, on the expense of the other. For $1.413 \leq t$ the gyroid surface is no longer connected and will be neglected in the following discussion.

Let us consider one of the two gyroid networks formed by the minority phase. Its surface can be described by Eq. 2.1. For $t = 0$, the single gyroid network occupies

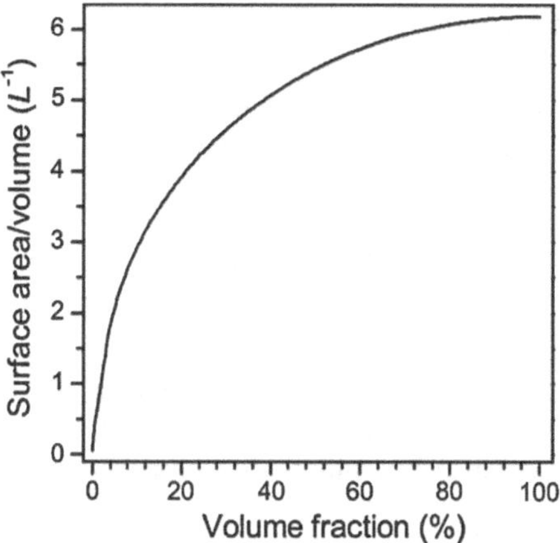

Fig. 2.8 Specific surface area to bulk volume ratio in units of the cubic unit cell dimension L of the double gyroid plotted versus the filling fraction. Adapted from [19]

exactly one half of the total volume, i.e. it has filling fraction $f = 50\,\%$. For $t > 0$ ($t < 0$) the filling fraction decreases (increases) monotonically and the surface of the gyroid network has no longer zero but an increasing (decreasing) constant mean curvature.

The surface area of the gyroid is related to its filling fraction. For the double gyroids the surface area per unit volume is plotted in Fig. 2.8. In the single gyroid the relation of surface area and filling fraction is analogous, with a maximum for $f = 50\,\%$ and a symmetric decaying for $f > 50\,\%$.

2.1.3 Block Copolymer Applications

Block copolymers have received much attention in the past two decades with increasing applications in nanotechnology. Nonetheless, their commercial interest has started over 50 years ago. Block copolymers can be found in products as shoe soles, tire treads, adhesive tape and asphalt additives that combine high-temperature resilience and low-temperature flexibility. The importance of block copolymers can be seen in their wide array of combined properties made possible by different conjuncted polymers.

The applications in nanotechnology are related to the scale of the microphase separation and to the tunability of the microphases, regulated simply by the molecular weight and composition, and to the wide range of chemical and physical properties of the constituent polymer blocks.

Block copolymer films have been used principally in nanolithographic etching processes, where the polymer phases function as a mask with high density of features [20]. They have also been utilized to develop periodic photonic band gap materials.

The first example was a simple 1D photonic crystal made with high molecular weight lamellae [21].

Other block copolymer applications in nanotechnology include porous membranes based on the large surface/volume ratio, low density and pore size monodispersity [22].

The replication of one or more phases to use a block copolymer as sacrificial template has been widely demonstrated. Metals are typically deposited by electroless deposition, while oxides and other dielectrics are produced by atomic layer deposition (ALD) or sol-gel chemistry [23].

The replication of the gyroid morphology into metals was also achieved by electroless deposition of nickel for catalytic applications [24] and by sol-gel chemistry to create a highly porous SiO_2 structure with a reduced refractive index.

Gyroid block copolymers have found applications in supercapacitors [25], solar cells [26] and nanoporous membranes [27], and represent to date a versatile structure for an increasing number of applications.

However, the main limitation of the block copolymers applications in nanotechnology lies with the fact that, while the self assembled structure is locally very precise, it is difficult to control its order on larger scales. The domain size of the periodic ordering are typically of several tens of the lattice constants, limiting the long-range order to 2–3 μm.

In the next chapter I will discuss a fabrication process that enables the ordering of the gyroid block copolymer into domains of several hundreds of microns, an essential requirement for engineering applications and optical characterizations.

2.2 Metamaterials

Metamaterials are artificially structured materials which are engineered to gain their properties not only from their composition, but from their design. Their geometry, size and arrangement can affect the waves of light or sound in a unconventional manner, giving rise to properties which are not available in any other bulk material (the word ‘meta’ means ‘beyond’ in Greek, and in this sense the name metamaterial refers to ‘beyond conventional materials’). The structural units have dimensions much smaller than the wavelengths of interest, so that the metamaterial is perceived homogeneous by the incident waves and their electromagnetic response is expressed in terms of homogenized material parameters.

Sought-after properties in metamaterials are the simultaneous negative electric permittivity ϵ and magnetic permeability μ. This leads to a negative refraction as demonstrated by Veselago in 1968 [28]. Negative index metamaterials have the potential to form superlenses, overcoming the diffraction limit, or to enable novel optical effects, including cloaking [29, 30].

Although negative permittivity is quite common in metals at optical wavelengths, it is very unusual to find natural materials that exhibit a magnetic response at terahertz or higher frequencies [31].

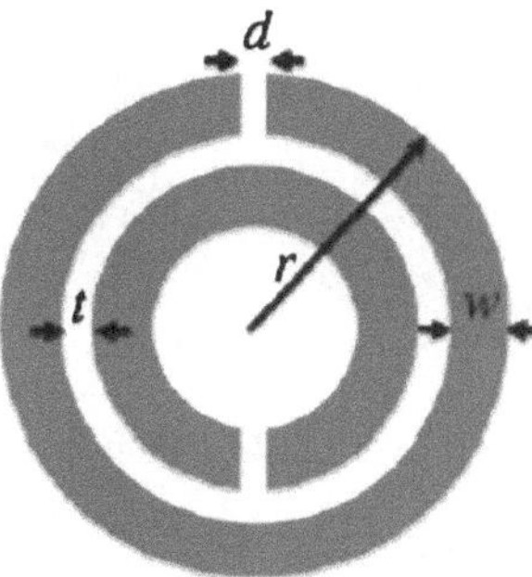

Fig. 2.9 Split-ring structure proposed by Pendry et al. [32]

However, a metamaterial made of a conductive but not-magnetic material designed as a split-ring resonator (SRR) can display a magnetic response and negative permeability at the resonance frequencies of this oscillator, as first predicted by John Pendry in 1999 [32]. A single cell SRR consist of a pair of enclosed loops with splits in opposite sides as shown in Fig. 2.9. They represent a paradigm example of a metamaterial system. A magnetic wave penetrating the metal rings produces an oscillating current flow that induces local magnetic dipole moments giving an effective magnetic response and thereby an engineered permeability.

When a SRR array is combined with an array of conducting wires it can create a medium with simultaneous negative permeability a permittivity.

Smith and coworkers showed evidence of negative index refraction with ring resonators in 2001 [33]. The first metamaterials were demonstrated at microwave frequencies and then scaled to the mid-infrared [34]. However, further down-scaling requires a different approach as a consequence of the strong deviation of metals from ideal conductors at high frequencies [35]. Moreover, to extend metamaterials to optical frequencies requires scaling the constituting features from the millimetre size down to the nanometre scale.

Several metamaterial architectures have been proposed in the past decade, some of which are summarized in Fig. 2.10. The operating frequency of optical metamaterials have already reached visible wavelengths, and negative refraction has been proved for light wavelengths between 720 and 760 nm [36].

However, some critical problems still remain unsolved. The main drawback for most potential applications of metamaterials are the high losses originated from the metal in the structure. These are a critical problem mainly at the infrared and optical frequencies where the metals differs from a perfect conductor. This results in high absorption and strong reduction of light transmission through the metamaterials.

Another problem in the fabrication of metamaterials is the development of truly 3D structures. The current metamaterials are generally limited to light coming from a narrow range of directions.

The proposed block copolymer based metamaterial is truly three dimensional and, as we will see in the following chapters, it displays strongly enhanced transmission,

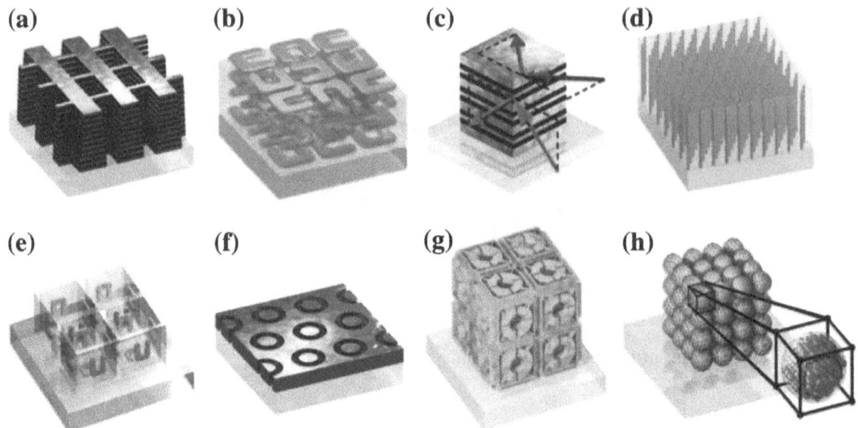

Fig. 2.10 Principal metamaterial designs proposed in the past decade. **a** Fishnet structure [37]. **b** Stereo metamaterial [38]. **c** Hyperbolic metamaterial [39]. **d** Metal-dielectric layered metamaterial [40]. **e** Three dimensionally oriented SRRs [41]. **f** Coaxial metamaterial [42]. **g** Connected cubic-symmetry metamaterial [43]. **h** Metal cluster-of-clusters array [44]. Adapted from [45]

i.e. it has low losses. Moreover, it is produced by self-assembly in contrast to the largely used top-down techniques of current metamaterials.

2.2.1 Chiral Metamaterial

It was recently proposed [46] that a chiral structure can offer a unique route to a negative refractive index for one circular polarization. A structure is defined as chiral if it lacks any planes of mirror symmetry.

The existence of chirality breaks the degeneracy of two oppositely circularly polarized waves. The two polarized beams experience different propagation speeds in a chiral media, i.e. the refractive index is increased for one circular polarization and reduced for the other. Therefore, if the chirality is strong enough, negative refraction may occur for one circularly polarized wave even when permittivity and permeability are both positive.

The first chiral metamaterial was proposed by Pendry in his pioneering work [46] as a "Swiss Roll" structure. Other theoretical works have also demonstrated negative refraction in the Thz and microwave regimes [47]. The principal experimentally obtained chiral metamaterials are summarized in Fig. 2.11.

The complex structure of chiral metamaterials has been a significant challenge in the miniaturization to the nanometric scale. Standard top-down approaches such as lithography techniques are not ideal to fabricate complex three dimensional chiral structures whereas the self-assembly of gyroid block copolymer is an ideal candidate to produce a chiral metamaterial.

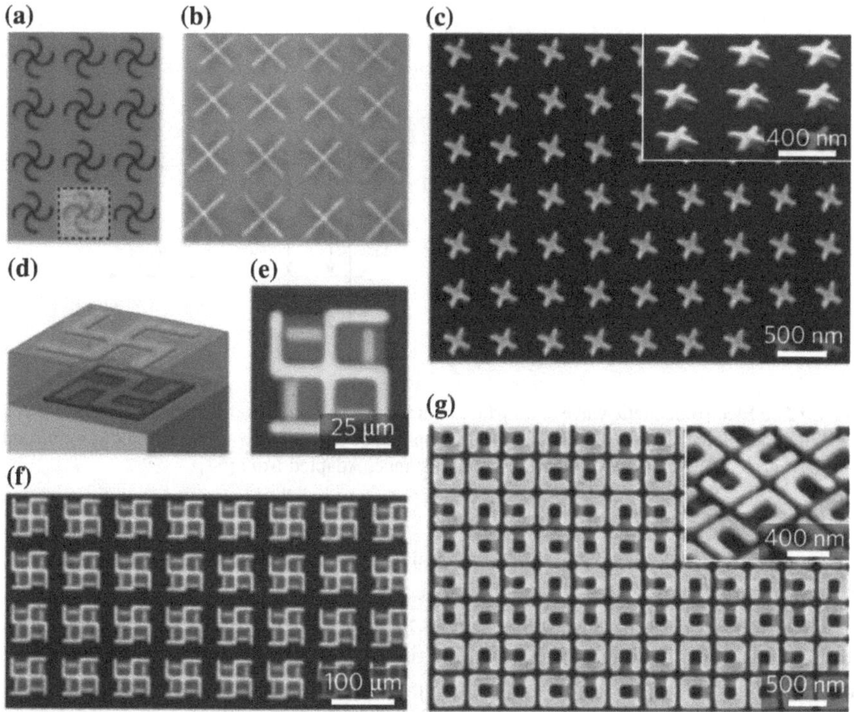

Fig. 2.11 Chiral metamaterial structures proposed in the past years. **a** Rosette chiral metamaterial with negative refraction at gigahertz frequencies [48]. **b** Cross-wire chiral metamaterials with negative refraction at GHz frequencies [49]. **c** Chiral structure composed of *right-handed* twisted gold crosses [50]. **d–f** Tunable chiral metamaterial [51]. **g** Twisted *U-shaped* split ring resonators [52]. Adapted from [45]

2.2.2 Surface Plasmons

Metamaterials with features on the nanometre scale derive their optical properties from surface plasmon waves. These are collective oscillations of free electrons on the surface of metallic interfaces with dielectrics. The free electrons dominate the interaction of light with metals. The optical electromagnetic field produces a force that displaces these weakly bound electrons. Since the electrons are negatively charged particles in a background of positive charges, the electron movement produces a restoring force. The free electrons can be considered as optically driven oscillators with with a natural frequency of oscillation known as the plasma frequency. This is given by:

$$\omega_{\mathrm{p}} = \sqrt{\frac{n_{\mathrm{e}} e^2}{\epsilon_0 m^*}} \tag{2.2}$$

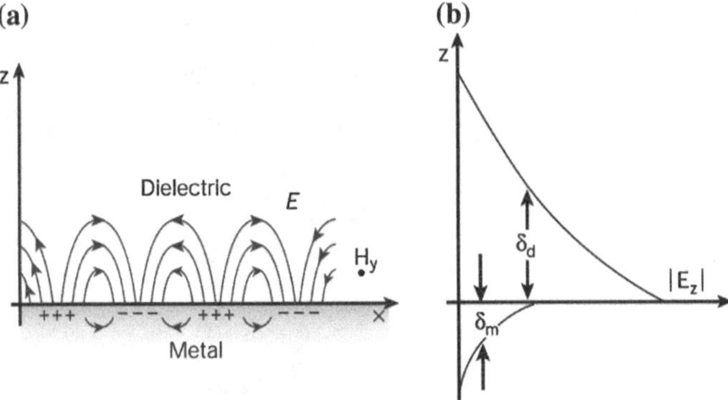

Fig. 2.12 **a** Electromagnetic wave and surface charges of surface plasmons at the interface between a metal and a dielectric material. **b** The field component perpendicular to the surface is enhanced near the surface and decays exponentially with distance. Adapted from [54]

where ϵ_0 is the dielectric constant of free space, e is the charge of the electrons, n_e is their number density, and m^* is their effective mass.

The interaction of the electromagnetic field and the surface free electrons produces surface plasmons that can propagate along the surface. In contrast to the propagating nature of the surface plasmons, the field perpendicular to the surface decays exponentially with distance from the surface. The field in this perpendicular direction is said to be evanescent. In the dielectric medium above the metal, typically air, the decay length of the field is on the order of one-half of the wavelength of light, whereas the decay length into the metal is determined by the skin depth, as represented in Fig. 2.12.

The skin depth of noble metals in the visible spectrum is on the order of 20 nm [53]. As the strut thickness of the gold gyroid is only 10 nm, the entire gyroid network acts effectively as a 'surface' and all the material is permeated by surface plasmons.

As we will see in the next chapters the effect of surface plasmons in the gyroid metamaterial enables enhanced transmission through gold gyroid films of several hundred nanometers.

2.2.3 Gyroid Theoretical Models

The gyroid morphology has received increasing attention as metamaterial structure, leading to theoretical predictions of its optical properties. However, these results are still debated.

Hur et al. [55] have calculated the photonic band structures of single and double gyroids made of gold and silver. The double gyroid morphology was predicted to form a capacitor, leading to a different light propagation mechanism than in the single gyroid morphology. The light propagation was suggested to originate from the coupled surface plasmons resonance on the closed loops formed in the double gyroid, supporting the surface plasmon propagation.

Negative refraction was predicted in silver and aluminium double gyroids, with n inversely proportional to the unit cell size and the frequency range controlled by the filling fraction. According to this study, the strong absorption deriving from the inter-band transitions impede the negative refraction in gold gyroids. Moreover, in the single gyroid, the lack of a counter electrode was predicted to inhibit light propagation through the metallic gyroid network.

Oh et al. [10] simulated the gyroid metamaterials by approximating the gyroid with three orthogonal helices. They identified one longitudinal and two degenerate transverse modes in the gold single gyroid. Their results suggest that the chirality may be smaller than expected as a direct consequence of the mixture of right- and left-handed helices shown in Fig. 2.4.

The only experimental work on gold gyroid metamaterials was presented by Vignolini et al. in 2011 [56]. It showed a strong reduction of the gold plasma frequency and circular dichroism. These results will be discussed more in detail in the next chapters.

References

1. Kratochvil P (1996) Glossary of basic terms in polymer science. Pure Appl Chem 68:2287–2311
2. Hamley IW et al (1998) The physics of block copolymers, volume 19. Oxford University Press, New York
3. Webster OW (1991) Living polymerization methods. Science 251(4996):887–893
4. Flory PJ (1942) Thermodynamics of high polymer solutions. J Chem Phys 10:51–61
5. Bates FS, Fredrickson GH (1999) Block copolymers—designer soft materials. Phys Today 52(2):32–38
6. Wu Y, Cheng G, Katsov K, Sides SW, Wang J, Tang J, Fredrickson GH, Moskovits M, Stucky GD (2004) Composite mesostructures by nano-confinement. Nat Mater 3(11):816–822
7. Nagpal U, Detcheverry FA, Nealey PF, de Pablo JJ (2011) Morphologies of linear triblock copolymers from monte carlo simulations. Macromolecules 44(13):5490–5497
8. Shefelbine TA, Vigild ME, Matsen MW, Hajduk DA, Hillmyer MA, Cussler EL, Bates FS (1999) Core-shell gyroid morphology in a poly (isoprene-block-styrene-block-dimethylsiloxane) triblock copolymer. J Am Chem Soc 121(37):8457–8465
9. Zheng W, Wang Z-G (1995) Morphology of abc triblock copolymers. Macromolecules 28(21):7215–7223
10. Oh SS, Demetriadou A, Wuestner S, Hess O (2012) On the origin of chirality in nanoplasmonic gyroid metamaterials. Adv Mater 25(4):612–617
11. Epps TH, Cochran EW, Bailey TS, Waletzko RS, Hardy CM, Bates FS (2004) Ordered network phases in linear poly(isoprene-b-styrene-b-ethylene oxide) triblock copolymers. Macromolecules 37(22):8325–8341
12. Schoen AH (1970) Infinite periodic minimal surfaces withouty self-intersections. Nasa technical note, D(5541)
13. Scherer MRJ (2013) Double-Gyroid-Structured functional materials: synthesis and applications. Springer Science & Business, New York
14. Große-Brauckmann K, Meinhard W (1996) The gyroid is embedded and has constant mean curvature companions. Calc Var Partial Differ Equ 4(6):499–523

15. Longley W, McIntosh TJ (1983) A bicontinuous tetrahedral structure in a liquid-crystalline lipid. Nature 303:612–614
16. Michielsen K, Stavenga DG (2008) Gyroid cuticular structures in butterfly wing scales: biological photonic crystals. J R Soc Interface 5(18):85–94
17. Saranathan V, Osuji CO, Mochrie SGJ, Noh H, Narayanan S, Sandy A, Dufresne E, Prun RO (2010) Structure, function, and self-assembly of single network gyroid (I4132) photonic crystals in butterfly wing scales. PNAS 107(26):11676–11681
18. Hyde S, Blum Z, Landh T, Lidin S, Ninham BW, Andersson S, Larsson K (1996) The language of shape: the role of curvature in condensed matter: physics, chemistry and biology. Elsevier, The Netherlands
19. Scherer MRJ, Li L, Cunha PMS, Scherman OA, Steiner U (2012) Enhanced electrochromism in Gyroid-Structured Vanadium Pentoxide. Adv Mater Deerfield Beach Fla 24(9):1217–1221
20. Segalman RA (2005) Patterning with block copolymer thin films. Mater Sci Eng: R: Rep 48(6):191–226
21. Edrington AC, Urbas AM, DeRege P, Chen CX, Swager TM, Hadjichristidis N, Xenidou M, Fetters LJ, Joannopoulos JD, Fink Y et al (2001) Polymer-based photonic crystals. Adv Mater 13(6):421–425
22. Hashimoto T, Tsutsumi K, Funaki Y (1997) Nanoprocessing based on bicontinuous microdomains of block copolymers: nanochannels coated with metals. Langmuir 13(26):6869–6872
23. Park C, Yoon J, Thomas EL (2003) Enabling nanotechnology with self assembled block copolymer patterns. Polymer 44(22):6725–6760
24. Hsueh H-Y, Huang Y-C, Ho R-M, Lai C-H, Makida T, Hasegawa H (2011) Nanoporous gyroid nickel from block copolymer templates via electroless plating. Advanced Materials (Deerfield Beach, Fla.) 23(27):3041–3046
25. Wei D, Scherer MRJ, Bower C, Andrew P, Ryhanen T, Steiner U (2012) A nanostructured electrochromic supercapacitor. Nano Lett 12(4):1857–1862
26. Crossland EJW, Kamperman M, Nedelcu M, Ducati C, Wiesner U, Smilgies D-M, Toombes GES, Hillmyer MA, Ludwigs S, Steiner U et al (2008) A bicontinuous double gyroid hybrid solar cell. Nano Lett 9(8):2807–2812
27. Li L, Schulte L, Clausen LD, Hansen KM, Jonsson GE, Ndoni S (2011) Gyroid nanoporous membranes with tunable permeability. ACS Nano 5(10):7754–7766
28. Veselago VG (1968) The electrodynamics of substances with simultaneously negative values of epsilon and mu. Sov Phys Uspekhi 10(4):509–514
29. Pendry JB (2000) Negative refraction makes a perfect lens. Phys Rev Lett 85(18):3966–3969
30. Pendry JB, Schurig D, Smith DR (2006) Controlling electromagnetic fields. Science (New York, N.Y.) 312(5781):1780–1782
31. Yen T-J, Padilla WJ, Fang N, Vier DC, Smith DR, Pendry JB, Basov DN, Zhang X (2004) Terahertz magnetic response from artificial materials. Science 303(5663):1494–1496
32. Pendry JB, Holden AJ, Robbins DJ, Stewart WJ (1999) Magnetism from conductors and enhanced nonlinear phenomena. IEEE Trans Microw Theory Tech 47(11):2075–2084
33. Smith DR, Padilla WJ, Vier DC, Nemat-Nasser SC, Schultz S (2000) Composite medium with simultaneously negative permeability and permittivity. Phys Rev Lett 84(18):4184–4187
34. Wu W, Yu Z, Wang S-Y, Williams RS, Liu Y, Sun C, Zhang X, Kim E, Shen YR, Fang NX (2007) Midinfrared metamaterials fabricated by nanoimprint lithography. Appl Phys Lett 90(6):063107–063107
35. Zhou J, Koschny T, Kafesaki M, Economou EN, Pendry JB, Soukoulis CM (2005) Saturation of the magnetic response of split-ring resonators at optical frequencies. Phys Rev Lett 95(22):223902
36. Xiao S, Drachev VP, Kildishev AV, Ni X, Chettiar UK, Yuan H-K, Shalaev VM (2010) Loss-free and active optical negative-index metamaterials. Nature 466(7307):735–738
37. Zhang S, Fan W, Panoiu NC, Malloy KJ, Osgood RM, Brueck SR (2006) Optical negative-index bulk metamaterials consisting of 2d perforated metal-dielectric stacks. Opt Express 14(15):6778–6787

38. Katsarakis N, Konstantinidis G, Kostopoulos A, Penciu RS, Gundogdu TF, Kafesaki M, Economou EN, Koschny T, Soukoulis CM (2005) Magnetic response of split-ring resonators in the far-infrared frequency regime. Opt Lett 30(11):1348–1350
39. Fang A, Koschny T, Soukoulis CM (2009) Optical anisotropic metamaterials negative refraction and focusing. Phys Rev B 79(24):245127
40. Hoffman AJ, Alekseyev L, Howard SS, Franz KJ, Wasserman D, Podolskiy VA, Narimanov EE, Sivco DL, Gmachl C (2007) Negative refraction in semiconductor metamaterials. Nat Mater 6(12):946–950
41. Burckel DB, Wendt JR, Eyck GAT, Ginn JC, Ellis AR, Brener I, Sinclair MB (2010) Micrometer-scale cubic unit cell 3d metamaterial layers. Adv Mater 22(44):5053–5057
42. Burgos SP, de Waele R, Polman A, Atwater HA (2010) A single-layer wide-angle negative-index metamaterial at visible frequencies. Nat Mater 9(5):407–412
43. Güney DO, Koschny T, Soukoulis CM (2010) Intra-connected three-dimensionally isotropic bulk negative index photonic metamaterial. Opt Express 18(12):12348–12353
44. Rockstuhl C, Lederer F, Etrich C, Pertsch T, Scharf T (2007) Design of an artificial three-dimensional composite metamaterial with magnetic resonances in the visible range of the electromagnetic spectrum. Phys Rev Lett 99(017401):1–4
45. Soukoulis CM, Wegener M (2011) Past achievements and future challenges in the development of three-dimensional photonic metamaterials. Nat Photonics 5:523–530
46. Pendry JB (2004) A chiral route to negative refraction. Science (New York, N.Y.) 306(5700):1353–1355
47. Zhao R, Koschny T, Soukoulis CM (2010) Chiral metamaterials: retrieval of the effective parameters with and without substrate. Opt Express 18(14):14553–14567
48. Plum E, Zhou J, Dong J, Fedotov VA, Koschny T, Soukoulis CM, Zheludev NI (2009) Metamaterial with negative index due to chirality. Phys Rev B 79(035407):1–6
49. Zhou J, Dong J, Koschny T, Kafesaki M, Soukoulis CM (2009) Negative refractive index due to chirality. arXiv preprint arXiv:0907.1121
50. Decker M, Ruther M, Kriegler CE, Zhou J, Soukoulis CM, Linden S, Wegener M (2009) Strong optical activity from twisted-cross photonic metamaterials. Opt Lett 34(16):2501–2503
51. Zhou J, Taylor A, O'Hara J, Chowdhury R, Zhao R, Soukoullis CM (2010) Chiral thz metamaterial with tunable optical activity. Technical report, Los Alamos National Laboratory (LANL), Los Alamos
52. Decker M, Zhao R, Soukoulis CM, Linden S, Wegener M (2010) Twisted split-ring-resonator photonic metamaterial with huge optical activity. Opt Lett 35(10):1593–1595
53. Hajiaboli A, Kahrizi M, Truong V-V (2012) Optical behaviour of thick gold and silver films with periodic circular nanohole arrays. J Phys D: Appl Phys 45(48):485105
54. Barnes WL, Dereux A, Ebbesen TW (2003) Surface plasmon subwavelength optics. Nature 424(6950):824–830
55. Hur K, Francescato Y, Giannini V, Maier SA, Hennig RG, Wiesner U (2011) Three-dimensionally isotropic negative refractive index materials from block copolymer self-assembled chiral gyroid networks. Angew Chem(International ed. in English) 50(50):11985–11989
56. Vignolini S, Yufa NA, Cunha PS, Guldin S, Rushkin I, Stefik M, Hur K, Wiesner U, Baumberg JJ, Steiner U (2012) A 3D optical metamaterial made by self-assembly. Adv Mater (Deerfield Beach, Fla.) 24(10):OP23-7

Chapter 3
Gyroid Metamaterial Fabrication

3.1 Introduction

The basic steps to produce an optical metamaterial from block copolymer self-assembly can be summarized in the following: (a) self-assembly of gyroid block copolymer in thin films; (b) selective degradation and removal of one of the minor phases; (c) back filling of the so created nanoporous template via electrodeposition; (d) removal of the remaining polymer template. Figure 3.1 shows a schematic representation of the gold gyroid fabrication in the case of single gyroid from a triblock copolymer.

The self-assembly of the gyroid triblock copolymer in thin films and the replication of one of the minor phases into gold was accomplished by Nataliya Yufa in 2009. Although this proof of principle enabled the first characterization of the gyroid gold metamaterial [1], the fabrication method was lacking in reproducibility and required a better understanding of the annealing procedure and substrate preparation.

In this chapter I will describe different annealing procedures that I have established and further developments in the fabrication process.

3.2 Substrate Preparation

The samples were prepared on 2.2 mm thick glass substrates, coated on one side with a 300 nm transparent conductive layer of fluorine tin oxide (FTO) to allow the optical characterization in transmission. The substrates were purchased from Solaronix SA (Catalogue No. TCO22-15 43273).

A crucial requirement for the metal replication is the continuity of the gyroid morphology at the bottom and top interfaces. In thin films, the continuity can be affected by asymmetric interaction energies with free or confining interfaces that can lead to surface reconstruction morphologies [2, 3]. Depending on the affinity between the substrate and each of the blocks, one the polymers might be favoured to contact the

© Springer International Publishing Switzerland 2015

S. Salvatore, *Optical Metamaterials by Block Copolymer Self-Assembly*, Springer Theses,
DOI 10.1007/978-3-319-05332-5_3

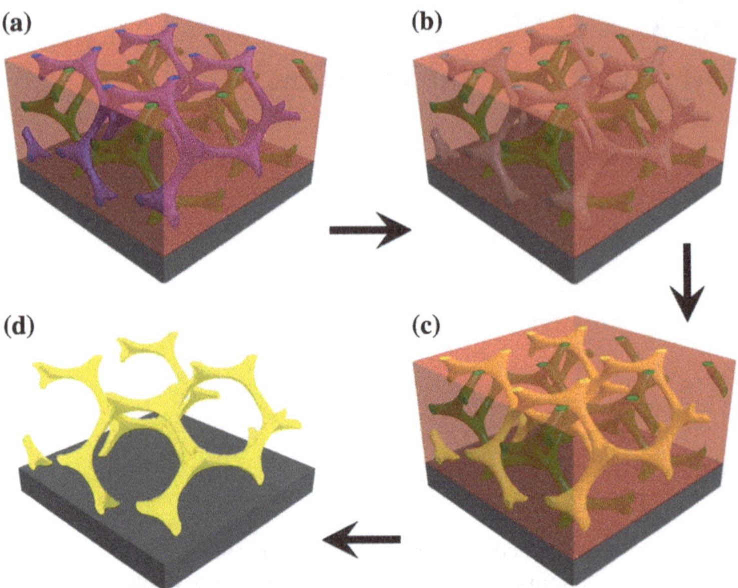

Fig. 3.1 Schematic representation of gold single gyroid fabrication. **a** The three differen colours (*red, blue,* and *green*) correspond to the different blocks of the copolymer. The isoprene block (*blue*) is removed (**b**) and is then back-filled with gold (**c**). The final 3D continuous gold network is revealed by removal of the two remaining polymer blocks by plasma etching (**d**)

substrate, resulting in the formation of a thin continuous layer underneath. To prevent such a reconstruction, the substrate was functionalized to increase the hydrophobicity by a procedure optimized for each substrate-block copolymer combination.

The FTO glass was initially cleaned for 20 min at 80 °C in a piranha solution formed by three parts in volume of sulfuric acid and one part of hydrogen peroxide. The substrates were then rinsed and sonicated for 10 min with deionised water to remove any residual piranha solution. They were then dried with a nitrogen gun and placed on a hotplate at 60 °C for 10 min to let the water evaporate. The substrates were then functionalized with silane to create a hydrophobic surface. In the case of ISO triblock copolymers, the silane functionalization was optimized at 20 s immersion in a solution of 0.2 %vol of octyltrichlorosilane in cyclohexane, followed by nitrogen flow drying.

3.3 I-S-O Triblock Copolymer

The triblock copolymers used in this project were synthesized and supplied by Wiesners group from Cornell University. They are terpolymers composed of polyisoprene (PI), polystyrene (PS) and polyethylene oxide (PEO), commonly named ISO block copolymer. The volume fraction of the three blocks was, respectively, 31, 53 and 16 % adopting the double gyroid morphology at equilibrium, with PI and PEO as the two minor single-gyroid phases embedded in the majority PS phase.

Two different batches were supplied with total molecular weights of 33 and 53 kg·mol^{-1} corresponding to measured unit cell sizes of 35 and 50 nm, respectively. The block copolymer were supplied as a powder in an amorphous state.

After the annealing and the self-assembly the PI phase was selectively removed by UV radiation (at 254 nm for 2 h) to break the carbon-carbon double bonds of the polyisoprene. The film was then immersed in ethanol to dissolve the cleaved polymer. While breaking the PI bonds the UV light induced also the cross-linking of the PS. After the metal electrodeposition the PS and PEO were removed by plasma etching.

3.3.1 Thermal Annealing

A solution of 10 wt% of ISO block copolymer in anisole was spin coated onto a FTO glass substrate to produce a film of about 1 µm thickness. During the spin-coating, the solvent was rapidly removed by centrifugal expulsion, viscous outflow and evaporation, leaving the block copolymer in a meta-stable amorphous state with no ordered phase separation.

Despite of the low affinity of the different polymers their mobility was not high enough to lead to the self-assembly at room temperature. To promote the self-assembly to the more energetically favourable gyroid phase the mobility of the polymer chains could be increased thermally.

The glass transitions of the polyisoprene and polyethylene oxide homopolymers are −70 and −54 °C respectively, whereas that of the polystyrene homopolymer is around 100 °C [4]. Therefore, above this temperature the block copolymer started behaving as a melt. The most effective thermal annealing resulted to be in vacuum at 180 °C for 30 min with heating rate of 150 °C/h followed by slow cooling in vacuum. The heating rate did not influence the process significantly whereas the cooling rate influenced the top surface reconstruction and the presence of the polyisoprene block at the top interface.

The thermal annealing produced very uniform samples with high reproducibility, but the size of the gyroid domains was limited to few a microns which represented a drawback for some optical characterization methods. The origin of the domain size can be identified considering the self-assembly mechanism. As the chain mobility is increased thermally, the amorphous film starts to self-assemble into the gyroid phase by a mechanism resembling nucleation and growth [5]. This process starts simultaneously across the entire film at distances on the same order of magnitude of the final size of the domains and terminates when the domain boundaries of the gyroid phases reach each other. A typical scale of the domains formed by thermal annealing are shown if Fig. 3.2. The increase of the domain size has been attempted by slowing down the heating rate in order to promote the growth process respect to

Fig. 3.2 Domain morphology of gyroids obtained by thermal annealing. The image is obtained after the metal replication by optical microscopy and under linear polarized light. As discussed in the next chapter, the metal gyroids show a strong linear dichroism that enables the inspection of the domain size by optical microscopy

the nucleations. No sensible domain increase was found and below 20 °C/min the long times at high temperature caused the degradation of the polymers.

For the optical properties discussed in the next chapters it is important to highlight that all ISO gyroid domains in thin films had a [110] out of plane orientation and random in-plane orientations (Fig. 3.3). This uniaxial orientation stems from the interaction of the block with the upper air surface of the film where nucleation starts in preferred orientation. Whether gyroid nucleation in thin films starts at the top or the bottom interface has been object of interest for a long time, because of the aim to induce domain ordering by substrate patterning [6].

To investigate the nucleation mechanism, the thin film was quenched during thermal annealing at an initial stage of the self-assembly process. The cross section of the film displayed in Fig. 3.4 clearly shows that the block copolymer had self-assembled only in the top part of the film, demonstrating that the self-assembly of the ISO block copolymer was nucleated at the top surface.

3.3.2 Drying-Annealing

The first gold gyroid metamaterials were produced in 2009 by Nataliys Yufa by drying annealing. A solution of 10 wt% of ISO block copolymer in anisole was blade coated onto the FTO glass and covered with a glass dome to slow down the solvent evaporation. The glass substrates were not functionalized and cleaned by acetone and isopropanol rinsing. Although this method produced some good samples for the first characterization, it lacked reproducibility.

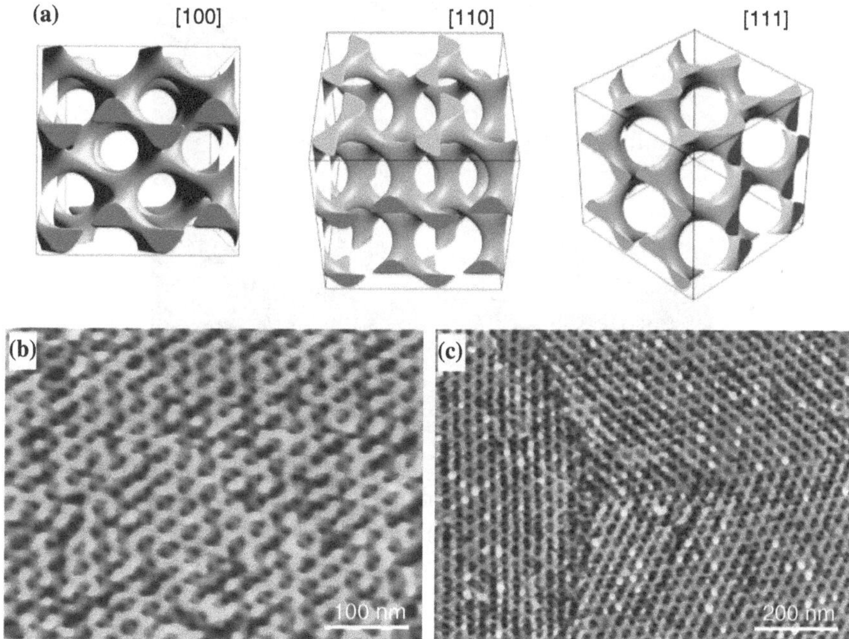

Fig. 3.3 **a** Schematic of the gyroid structure as seen from the indicated crystal directions. The [110] direction in the ISO triblock copolymer is always perpendicular to the surface of the gyroid samples as shown in (**b**), while the in plane orientation is completely random (**c**)

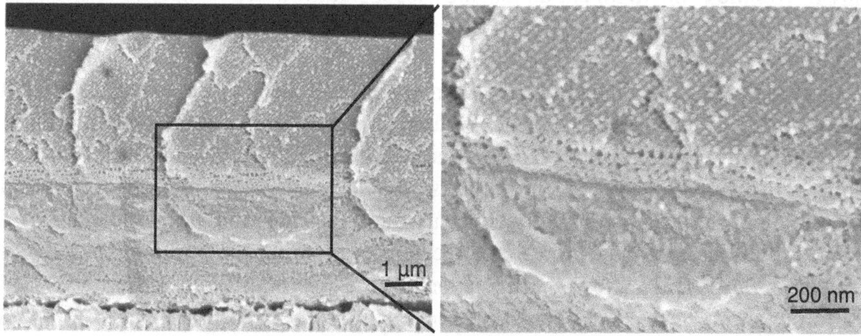

Fig. 3.4 Scanning electron microscope (SEM) image of the gyroid block copolymer cross section at the half-stage of the thermal annealing. The ordered structure at the top portion of the film suggest that the annealing process is initiated at the top interface

The procedure I developed allowed controlled dry annealing on a functionalized substrate. Normally, the drying annealing procedure is not suitable for hydrophobic substrate as the blade coated solution dewets, preventing thin film formation. On the other hand, a very concentrated and viscous solution cannot be used in blade coating. Nonetheless, a concentrated and viscous solution could be deposited onto the

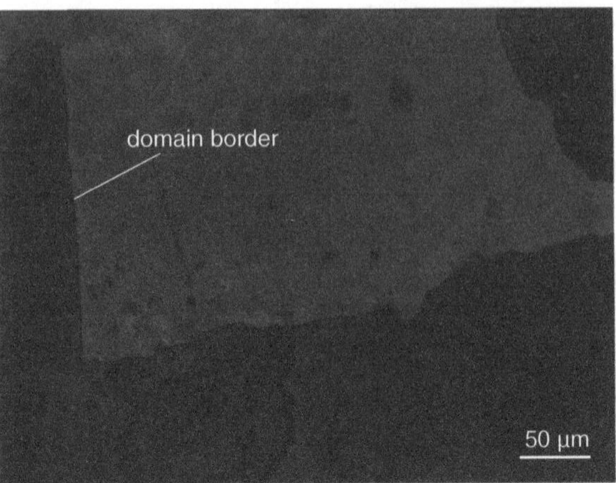

Fig. 3.5 Domain morphology of gyroids obtained by drying-annealing. The image was obtained by optical microscopy under linear polarized light after metal replication. As discussed in the next chapter, the metal gyroids show a strong linear dichroism that enables the inspection of the domain size

functionalized substrate by spin coating for very short times, forming a film. During the spin-coating most of the solution is rapidly removed by an initial centrifugal expulsion, followed by viscous outflow and finally by solvent evaporation. By spin coating the solution of 10 wt% ISO block copolymer in anisole at 1,200 rpm for only 8–10 s the centrifugal expulsion and viscous outflow are completed while only partial evaporation takes place, giving rise to a stable viscous thin film stable on the hydrophobic substrate.

During the drying annealing the polymers are initially dispersed in the solvent with high mobility and a reduced chain interaction. As the solvent slowly evaporates, the concentration of the solution increases. At a certain concentration the polymers, while still having some chain mobility, start microphase-separate into the gyroid morphology. As the solvent initially evaporates faster at the edges, leading to a drying front that propagates towards the centre of the film, this mechanism induces the self-assembly by a slow continuous process, which is different from nucleation and growth during thermal annealing. For this reason, by drying-annealing it was possible to obtain very large domains, on the order of hundreds microns as shown in Fig. 3.5.

3.3.3 *Solvent Vapour Annealing*

A more conventional solvent vapour annealing procedure was also investigated. The ISO block copolymer thin film was placed into a chamber with a solvent vapour control. The thin film interference of the block copolymers was monitored by a

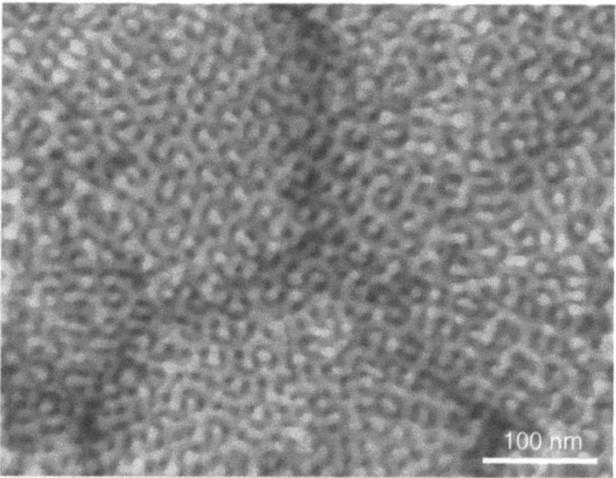

Fig. 3.6 SEM image of the gold double gyroid produced from the PS-PLA diblock copolymer

double fibre setup that allowed precise measurement of the film thickness related to the partial vapour pressure of the solvent. The anisole vapour pressure was increased to saturation and then slowly reduced to promote the gyroid self-assembly in a similar way as drying-annealing. Some successful gyroid formation was obtained for a decrease in solvent concentration below 4 %/h. However, the dimension of the domains of the gyroid obtained by solvent annealing were on the order of 5–10 μ, significantly smaller than those obtained by drying-annealing.

3.4 PFS-*b*-PLA Diblock Copolymer

Double gyroids were produced from a Poly(4-fluorostyrene)-*block*-poly(D,L-lactide) diblock copolymer (PFS-*b*-PLA, $M_w = 24\,\text{kgmol}^{-1}$ with 38 vol% PLA fraction) synthesized by Maik Scherer by atom transfer radical polymerization [7], followed by the addition of lactide via organocatalytic ring-opening polymerization [8].

In the PFS-*b*-PLA diblock copolymer the polylactide (PLA) formed two distinct gyroid networks embedded in the polyfluorostyrene (PFS) major phase. The thermal annealing procedure was developed by Maik Scherer. The polymer films were prepared by spin coating a 10 wt% toluene solution onto silanized FTO-glass substrates. The films were heated at 175 °C under a nitrogen atmosphere for 20 min at a ramping rate of 150 °C/hour. After quenching to room temperature, the films were soaked in 0.3 m NaOH solution containing 1:1 (wt) mixture of water and methanol for one hour to cleave the PLA phase [9, 10]. After electrodeposition the remaining polystyrene (PFS) majority phase was finally dissolved in toluene (Fig. 3.6).

Unlike the ISO triblock copolymer the PFS-*b*-PLA diblock copolymer produces a double gyroid with random orientation both in the plane and out of plane directions.

The size of the gyroid domains was only 1–2 μ after thermal annealing. In order to produce larger domains, drying-annealing was attempted using a similar procedure as for the ISO block copolymer. Unfortunately film dewetting occurred even at high solution concentrations, preventing the formation of large domains of the double gyroid.

3.5 Metal Deposition

After gyroid minor phase removal, the block copolymer templates were backfilled by metal electroplating. In an electroplating process the electric current reduces the metal cations dissolved in the electrolyte to metal, therby creating a deposit on the working electrode (i.e. the FTO layer). The electrochemical setup was composed of an FTO layer acting as working electrode, a saturated calomel electrode as reference electrode, and a platinum mesh as counter electrode. The electrodes were immersed in a liquid electrolyte containing the metal ions to be deposited. The three-electrode cell was controlled by a PGSTAT302N potentiostat and used in potentiostatic mode, holding the voltage between the working and the counter electrode at a defined value.

Metal electroplating follows a mechanism of nucleation and growth from the working electrode. To form continuous gyroid films within the block copolymer templates it was crucial to promote nucleation over the growth process. This was accomplished by cyclic voltammetry, sweeping the voltage from 0 to an over-potential that induced nucleation, and back to 0 again. The voltage was then set to a constant value for a time varied by the desired final thickness.

For gold deposition, 0.5 vol% brightener was added to the plating solution (Technic Inc., ECF60) to achieve a smoother deposition. The initial cyclic voltammetry ranged from 0 to −1.2 V–0 V at 50 mV/sec. The steady potential was set at −0.8 V.

For the nickel deposition, the plating solution was purchased from Alfa Aeser, the cyclic voltammetry voltage range was −1.4 V at 70 mV/sec and the steady potential was −1.1 V.

The silver plating solution was purchased from Metalor (Metsil 500 CNF), the cyclic voltammetry swept down to −1 V and the steady potential was held at −0.7 V.

3.6 Film Morphology Control

The thickness of the metal gyroid films produced by electrodeposition varied across the film due to the roughness of the FTO layer and to the fluctuations of the deposition rate. The resulting roughness was on the order of ±50 nm both at the top and bottom interfaces as shown in Fig. 3.7a–e. As discussed in the next chapter, some optical characterizations required gyroid films with very precise film thickness control, which stimulated the development of a new fabrication procedure.

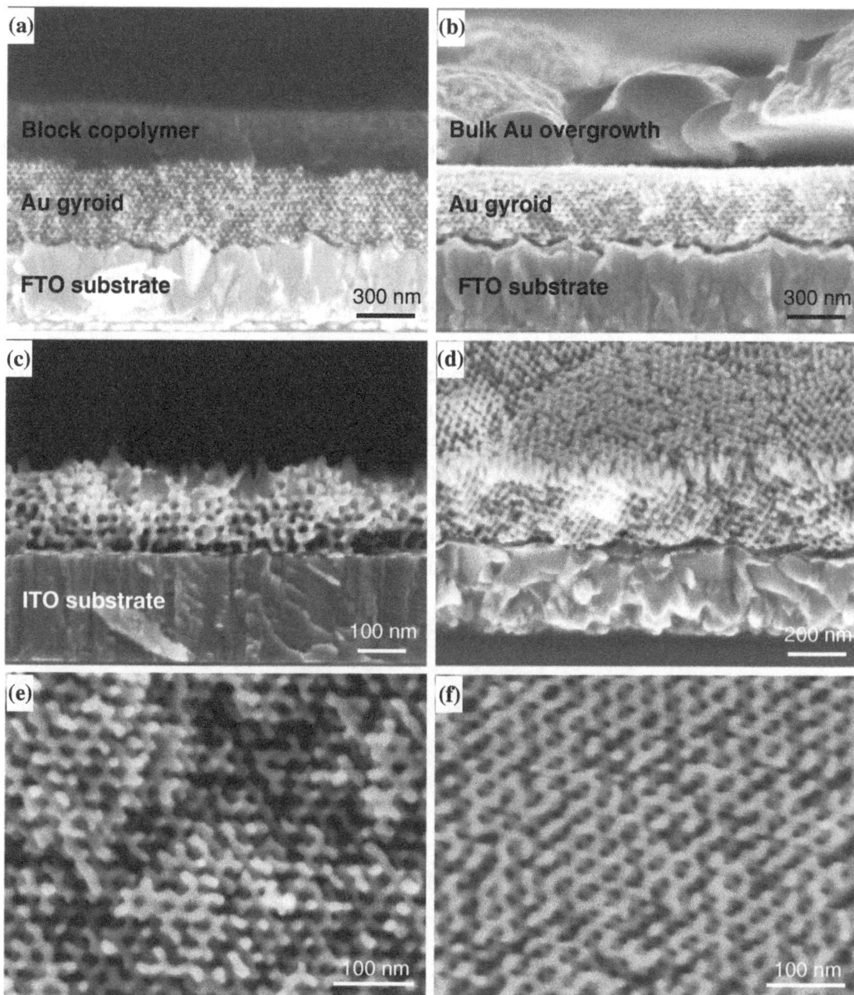

Fig. 3.7 **a** SEM image of the film cross-section revealing the roughness of the FTO substrate and of the gold gyroid electrodeposited into the block copolymer template. **b** SEM cross-sectional image of the metal gyroid produced by overgrowth removal. The thick layer of overgrowth gold can be detached from the film producing a flat surface of the metal gyroid. **c** SEM cross-sectional image of the gyroid produced on flat ITO substrate. **d** SEM image at 45° of gyroid produced on flat ITO substrate and by overgrowth removal. **e, f** SEM images in top view of the metal gyroid films produced without and with morphology control

The roughness of the bottom interface was reduced using glass coated with Indium Tin Oxide (ITO) as transparent conducting substrate. The FTO substrates are commonly preferred to ITO as the roughness is known to reduce the planar interface reconstruction of block copolymer self-assembly [3]. Gold was deposited into the voided self-assembeld ISO gyroid channels from the flat ITO glass (Fig. 3.7c) by functionalizing the substate by immersion in a solution of 0.5 vol% octyltrichlorosilane in cyclohexane for 30 s.

The roughness control at the top surface was achieved by developing a fabrication method termed overgrowth-removal. The metal was electrodeposited through the entire voided block copolymer template to form a thick overgrowth layer of bulk metal on top of the film. The continuous layer of bulk metal was then peeled off from the block copolymer template or cleaved by sonication, creating a smooth, planar surfaces of the metal electrodeposited, replicating the exact surface smoothness of the block copolymer film produced by spin coating (Fig. 3.7b–f).

Those fabrication methods could also be extended to produce gyroid films with design topography patterns, adding an extra degree of freedom to the design of the metamaterial structure. The proof of concept was realized by replicating a grating pattern into the gyroid film via nanoimprinting [11]. An optical grating was used as master and replicated in a polytetrafluoroethylene (PTFE) film at 230 °C and 40 bar. The patterned PTFE film was then used as master to imprint into the block copolymer film during the thermal annealing by applying a pressure of 20 bar. Finally, the metal replication of the block copolymer template successfully replicated both the gyroid morphology within the films and the surface topography of the grating as shown in Fig. 3.8.

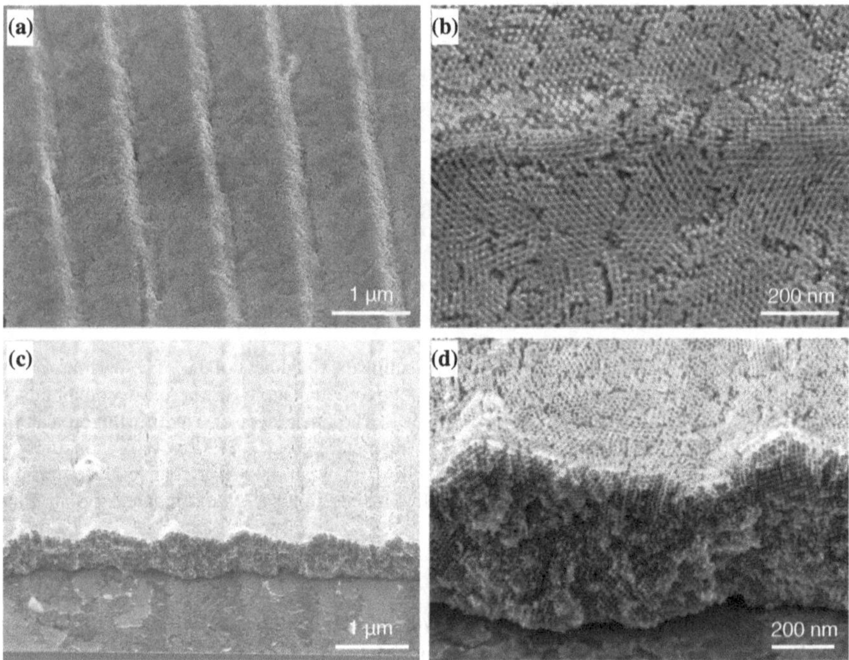

Fig. 3.8 SEM images of the metal gyroid with grating patterned topography produced by the overgrowth-removal process (**a**–**d**) details of the grating patterned structure

3.7 Conclusions

The good reproducibility achieved with the fabrication of gyroid metamaterial is an essential requirement for a consistent study of their optical properties. Moreover, it underlines the potential of the bottom-up approach by block copolymer self-assembly in producing a truly three dimensional optical metamaterial. The thermal annealing procedure gave rise to the highest quality uniformity and reproducibility, although the domain size was limited to only few microns. On the other hand, the drying annealing process enabled the fabrication of remarkably large domain sizes with a lower reproducibility, of about 30 %. In the following chapters I will present the optical properties of this gyroid metamaterials and some further developments in the fabrication process.

References

1. Vignolini S, Yufa NA, Cunha PS, Guldin S, Rushkin I, Stefik M, Hur K, Wiesner U, Baumberg JJ, Steiner U (2012) A 3D optical metamaterial made by self-assembly. Adv Mater (Deerfield Beach, Fla.) 24(10):OP23–OP27
2. Knoll A, Horvat A, Lyakhova K, Krausch G, Sevink G, Zvelindovsky A, Magerle R (2002) Phase behavior in thin films of cylinder-forming block copolymers. Phys Rev Lett 89(3):035501
3. Crossland EJW, Ludwigs S, Hillmyer MA, Steiner U (2010) Control of gyroid forming block copolymer templates: effects of an electric field and surface topography. Soft Matter 6(3):670
4. Wilkes CE (2005) PVC Handbook. Hanser, Munich
5. Naoki S, Takeji H (1998) Ordering dynamics of a symmetric polystyrene-block-polyisoprene. Real-space analysis on the formation of lamellar microdomain. Macromolecules 3(98):3815–3823
6. Segalman RA (2005) Patterning with block copolymer thin films. Mater Sci Eng R: Reports 48(6):191–226
7. Matyjaszewski K, Xia J (2001) Atom transfer radical polymerization. Chem Rev 101(9):2921–2990
8. Kamber NE, Jeong W, Waymouth RM, Pratt RC, Lohmeijer BGG, Hedrick JL (2007) Organocatalytic ring-opening polymerization. Chem Rev 107(12):5813–5840
9. Zalusky AS, Olayo-Valles R, Wolf JH, Hillmyer MA (2002) Ordered nanoporous polymers from polystyrene-polylactide block copolymers. J Am Chem Soc 124(43):12761–12773
10. Crossland EJW, Cunha P, Ludwigs S, Hillmyer MA, Steiner U (2011) In situ electrochemical monitoring of selective etching in ordered mesoporous block-copolymer templates. ACS Appl Mater Interface 3(5):1375–1379
11. Barbero DR, Saifullah MSM, Hoffmann P, Mathieu HJ, Anderson D, Jones GAC, Welland ME, Steiner U (2007) High-resolution nanoimprinting with a robust and reusable polymer mold. Adv Funct Mater 17(14):2419–2425

Chapter 4
Gyroid Metamaterial Characterization

4.1 Introduction

In this chapter I will present the full characterization of gyroid metamaterials.

The atomic structure of the gold gyroid was investigated using high resolution by transmission electron microscopy (TEM), revealing the small size of the metal grains and a very smooth surface of the gyroid struts.

The electrical characterization will then be discussed comparing the resistivity of the gyroid gold with the bulk gold.

Subsequently, an extensive section will be dedicated to the optical properties, studied by reflection and transmission analysis. A thin wire array model will be discussed to understand the origin of the optical response. Simulation spectra were produced in collaborations with two theoretical groups and will be presented together with my experimental results.

The larger part of the gyroid metamaterial investigation was conducted on a single gold gyroid, referred to simply as "gold gyroid", with a unit cell size of 35 nm and domains with the [110] orientation out of plane.

Finally, in the last part of the chapter I will discuss the optical properties of silver and double gyroids.

4.2 Transmission Electron Microscopy

The atomic structure of the gold gyroid was investigated by high resolution by transmission electron microscopy (TEM) [1] in collaboration with Dr. Caterina Ducati.[1] As depicted in Fig. 4.1 the gold atoms show crystalline structure with no lattice vacancies and the size of the grains is on the order of a few tens of nanometres. The dimension of the grains was influenced by the electrodeposition through

[1] Electron Microscopy Group, Department of Materials Science and Metallurgy, University of Cambridge.

© Springer International Publishing Switzerland 2015

S. Salvatore, *Optical Metamaterials by Block Copolymer Self-Assembly*, Springer Theses,
DOI 10.1007/978-3-319-05332-5_4

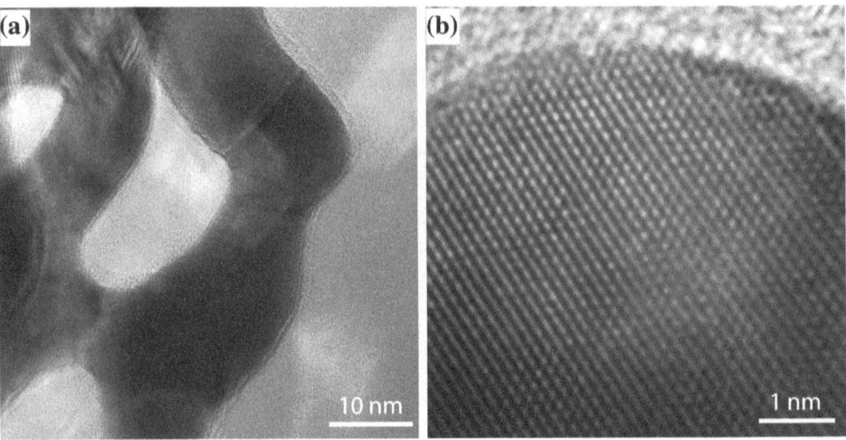

Fig. 4.1 Transmission electron microscopy (TEM) images of the gold gyroid. Different shades of *gray* in **a** represent the gold grains. **b** High resolution detail of a gyroid strut showing the metal crystallinity and very smooth surface

the nanosized network template that produces the formation of new grains at each struts intersection. The surface roughness of the gyroid struts is remarkably small, with imperfections only at the atomic scale.

Energy dispersive X ray diffraction did not reveal any other element in addition to gold, excluding contributions of impurities in the optical effects.

4.3 Electric Characterization

The nanoscale cross-section of the gyroid struts and the small dimension of the gold grains produced a significant increase of the electric resistivity as revealed by four probe measurements.

In the four probes setup, the probes were placed in one line, in contact with the gyroid film and a current was supplied to the two outer probes. The resistance of the gyroid film was extracted by the voltage drop between the two inner probes taking into account the geometrical factors of the probes. The resistivity of the supporting FTO layer from the product specifications was taken into account as a parallel resistor.

The four probe measurements yields the sheet resistance R_s, commonly defined as the resistance per aspect ratio. The intrinsic material resistivity ρ was then extracted by multiplying the R_s by the film thickness and the gold filling fraction. The measurement was performed on films with different thicknesses to verify the consistency of the measured values. The results are presented in Fig. 4.2 and revealed an increase of resistivity to $\rho_{\text{gyr}} \approx 52\,\mu\Omega\cdot\text{cm}$, about 20 times higher than of bulk gold $\rho_{\text{bulk}} \approx 2.44\,\mu\Omega\cdot\text{cm}$ [2].

Two contributions are responsible for the increase in the electric resistivity of gold in the gyroid architecture, the grain boundaries and surface scattering. From thin wire

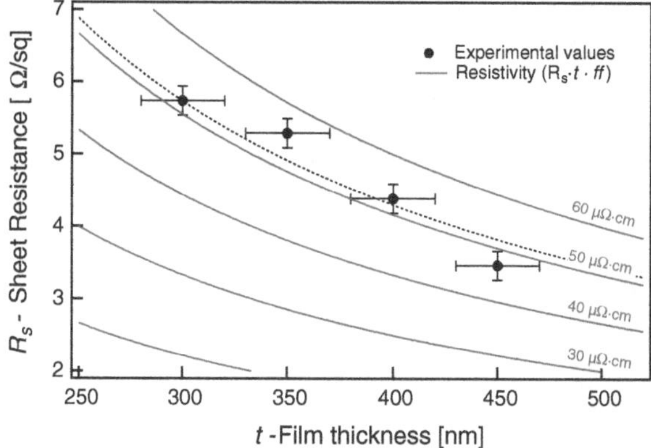

Fig. 4.2 Resistivity measurements for gold gyroid. The sheet resistance R_s of the gyroid film was measured by a four probe setup and multiplied by the film thickness and the filling fraction to extract the resistivity of gold in the nanostructured gyroid architecture. The values measured for different thicknesses reveal an average resistivity of ∼52 μΩ·cm, about 20 times higher than the value of 2.44 μΩ·cm for bulk gold. The *red lines* illustrate the variations of the resistivity in function of film thickness and sheet resistance

studies [3] it is know that for wires widths below approximately 0.5 times the mean grain size, surface scattering becomes important, approaching the same order of magnitude of grain-boundary scattering as the width decreases. Surface scattering and grain boundary imposes restrictions on the electron mean free path, which is 40 nm in gold [4].

As revealed by the TEM analysis, the grain size in the gold gyroid was 20–30 nm and the strut widthv was ∼10 nm. It is therefore reasonable to conclude that both grain boundary scattering and surface scattering have similar contributions to the resistivity.

Silver gyroids were also characterized electrically, resulting in a $\rho_{gyr} \approx 40\,\mu\,\Omega\cdot$cm. They had resistivities which were about 25 times higher than the bulk silver $\rho_{bulk} = 1.59\,\mu\Omega\cdot$cm.

4.4 Optical Properties

4.4.1 Single Gyroid

The optical response of the gold gyroid differs significantly from normal gold, exhibiting a red hue in reflection and a finite transmission (Fig. 4.3a, b).

The samples were characterized spectroscopically using an optical microscope with white light illumination and fiber detection in a confocal configuration. Confocal optical microscopy was performed using a 50 μm core optical fiber that served as pinhole in the conjugate to the focal plane of a ×20 microscope objective. A broadband halogen lamp acted as illumination source. Linear polarization measurements were obtained using achromatic polarizers (Thorlabs Inc.) and for circular polariza-

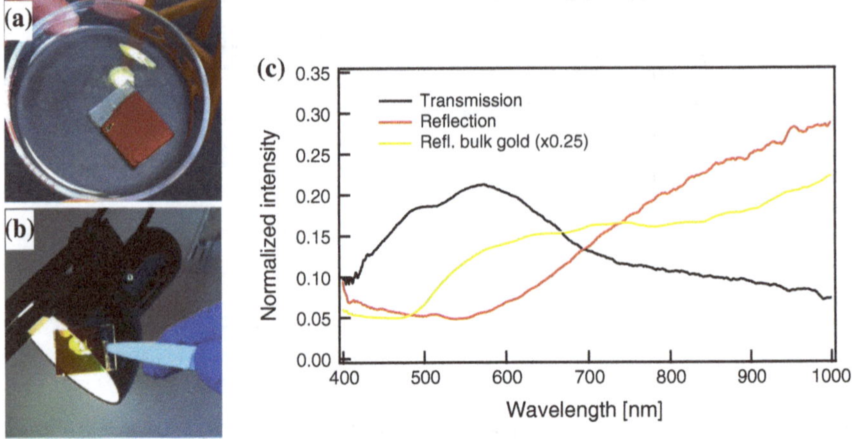

Fig. 4.3 **a, b** Appearance of the gold gyroid in reflection and transmission. **c** Transmission and reflection spectra for unpolarised incident light compared to bulk gold. Film thickness ~300 nm

tion measurements the polarizers were combined with superachromatic quarter wave quartz plates (B. Halle Nachfl. GmbH). Reflection measurements were normalized to a silver mirror reference, whereas transmission measurements were normalized to the FTO-glass substrate.

The experimental results for both reflection and transmission configurations are compared in Fig. 4.3c with the reflection of bulk gold produced by electrodeposition. The reflectivity of the gold gyroid is characterized by a dip at around 520 nm and a steady increase towards longer wavelengths.

A remarkable transmission of 20 % was found for a 300 nm thick layer with the smaller gyroid unit cell. This high transmission across a several-hundred-nanometer-thick layer with Au volume fractions of 30 % is evidence for the transport of the optical energy flux through the strut network by plasmon resonances which is the hallmark of an optical metamaterial.

4.4.1.1 Thin Wires Model

The optical properties of the gold gyroid can be understood considering in first approximation the gyroid in a simplified wire array model. Pendry et al. [5] showed that light interacting with an extended 3D network of thin wires (Fig. 4.4) induces a magnetic field that impedes the free electronic motion, producing an increase in the electron effective mass:

$$m_{\text{eff}} = \frac{\mu_0 e^2 \pi r^2 n}{2\pi} \ln(a/r), \tag{4.1}$$

where μ_0 is the permeability of free space, e the electron charge, n is the density of electrons in the wires, r the radius and a the cell side of the square lattice of the wires.

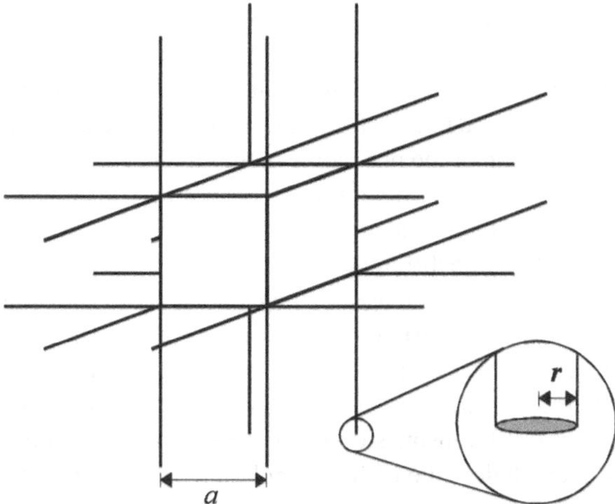

Fig. 4.4 Periodic structure model composed of infinite wires arranged in a cubic lattice. Adapted from Ref. [5]

The response of a material to electromagnetic radiation can be described by its dielectric function that relates the applied electric field to the local field created inside the material. In the Drude model, the dielectric function is derived only in terms of the response of the free electrons. These can be considered as classic oscillators with a characteristic oscillation frequency called the plasma frequency. The increase of the electron effective mass in the thin wire network modifies the plasma frequency and the damping of the electron motion by the relations:

$$\omega_{\text{eff}}^2 = \frac{n_{\text{eff}} e^2}{\epsilon_0 m_{\text{eff}}} \tag{4.2}$$

$$\Gamma_{\text{eff}} = \frac{\epsilon_0 \omega_{\text{eff}}^2 \rho}{f}, \tag{4.3}$$

where $n_{\text{eff}} = fn$ is the effective concentration of the metal free electrons modified by the filling fraction f, ϵ_0 is the vacuum permittivity, and ρ the resistivity of gold.

Therefore, the gyroid structure modifies the plasma frequency and the dielectric function of gold, effectively creating a new metal with its own optical properties.

The reflectivity of the gold gyroid arising from the effective dielectric function was simulated by Petros Farah[2] in transfer-matrix method simulations, modifying the Drude model to take into account the gold interband transitions, including the experimental value of the gold resistivity in the gyroid. The plasma frequency reduction was estimated from the bulk value of 7.5–2.05 eV, in good agreement to the experimental results.

[2] Nanophotonics group, University of Cambridge.

4.4.1.2 FDTD Calculations

Although the thin wire model was able to predict the reflection response of the gold gyroid in terms of the reduced plasma frequency and increased damping, it was unable to describe the propagation of the plasmonic modes in the gyroid. As a consequence, the transmission properties were not correctly described.

The simulation of the field enhancement and its propagation in the gold gyroid was developed by Oh[3] [6] by finite different time domain (FDTD) calculations [7], shown in Fig. 4.5. Both the electric and magnetic fields are less attenuated around 550 nm, in good agreement with the experimental transmission peak. The predicted magnetic field has a modest spacial variation around the network. In contrast, the electric field distribution is enhanced in the proximity of the gold struts. The absence of narrow peaks in the reflection and transmission spectra was an indication for the absence of localized modes. Nonetheless the FDTD calculations display the excitation of some weak localized surface plasmons.

4.4.1.3 Linear and Circular Dichroism

When illuminated under linearly polarized light the gold gyroid show strong birefringence. This effect is shown in Fig. 4.6 where two gyroid domains have in-plane orientation orthogonal to each other. Under unpolarized light illumination (Fig. 4.6a) the different domains are not distinguishable, whereas when the polarization of the incident light is parallel to the [110] in-plane direction of one of the areas and nearly perpendicular to the [110] in-plane direction of the neighbour, they show different optical responses.

This birefringence arises from the relative gyroid orientation and the polarization direction, where light is coupled differently to localized plasmon resonances. In Fig. 4.6d the variation of reflectivity is plotted as a function of the angle between the in-plane [110] gyroid direction and the incident light polarization. Stronger reflectivity was found for linear polarization perpendicular to the [110] in-plane orientation.

The gyrotropic behavior of the gyroid metamaterial was probed in a previous study by Vignolini et al. [8] by transmission measurements with left- and right-hand circularly polarized light. Figure 4.7 plots the gyrotropic tranmission as difference between the two circular polarizations, averaged over frequency region between 600 and 750 nm, for different gyroid orientations. As expected, the strongest gyrotropic effect was observed along the chiral [111] direction. As discussed in the previous chapter, the samples produced by the triblock copolymer gyroid exhibit constant [110] orientation in the out-of-plane direction. To access the [111] direction the sample was tilted by 35° and rotated around the [110] surface normal.

[3] Ortwin Hess' group, Department of Physics, Imperial College London.

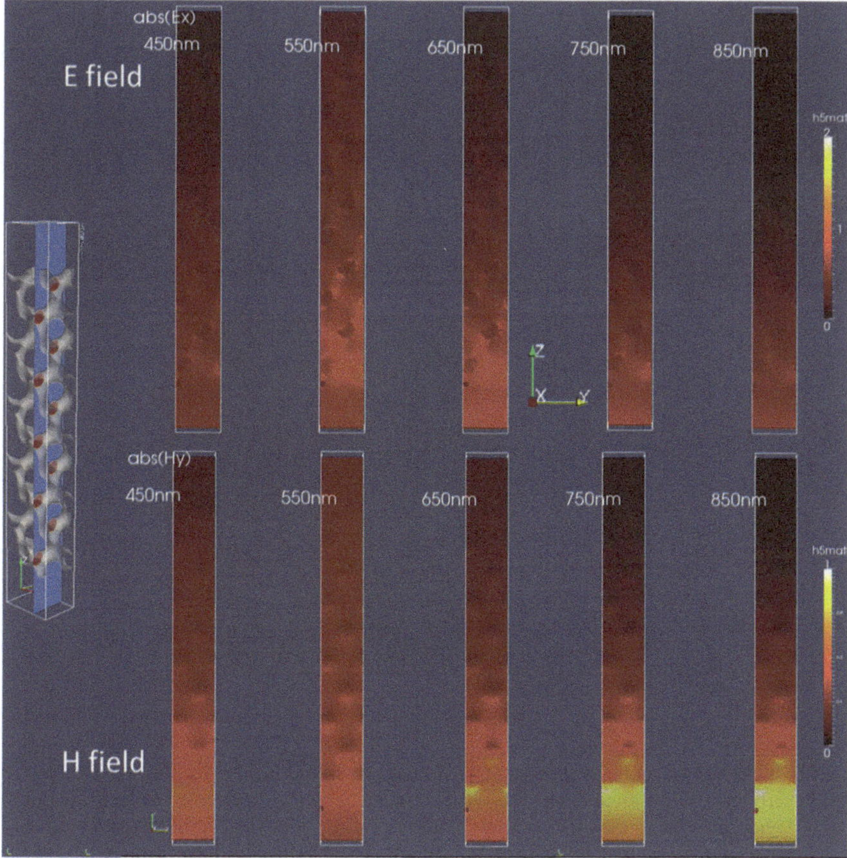

Fig. 4.5 FDTD calculation of the electric and magnetic field distributions in the metal gyroid for continuous wave excitation linearly polarized in the x direction. The magnetic field has a modest spacial variation around the gyroid, whereas the electric field distrubution is enhanced in proximity of the gyroid struts. Courtesy of Oh et al.

4.4.2 Film Thickness Effects

The film thickness effect was investigated by producing samples with thickness variation from 220–450 nm. The layer thickness was measured by cleaving the samples and examining the cross-sectioned edge by Scanning Electron Microscope (SEM).

Small fluctuations in the reflectivity response were observed from 300–450 nm as displayed in Fig. 4.8. The thinner film, of 220 nm, showed instead a very different response suggesting that a minimum number of unit cell was required to produce sufficient light absorption and self-inductance.

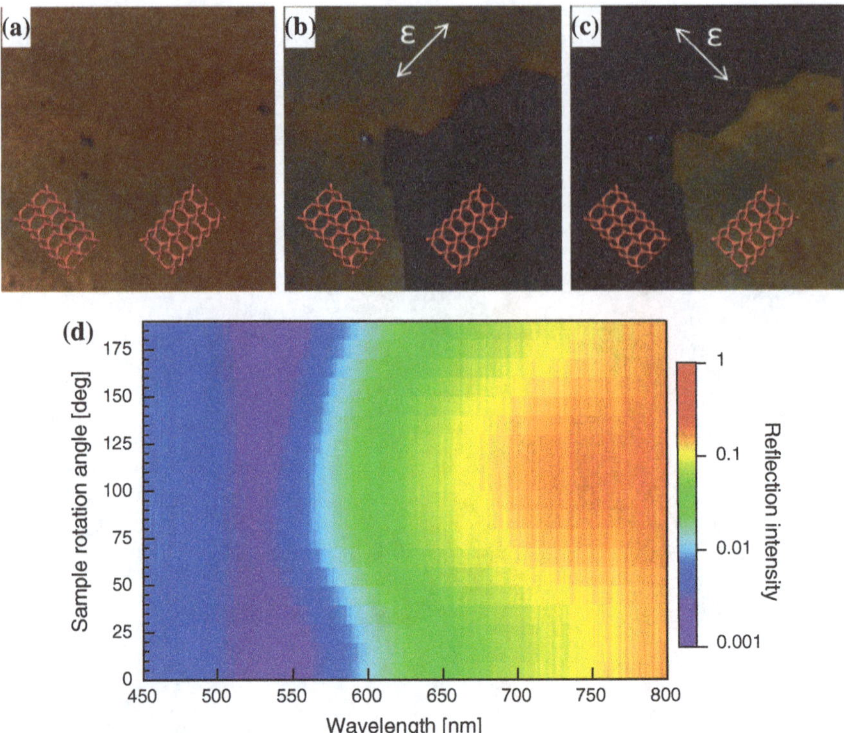

Fig. 4.6 **a** Optical reflection images for unpolarized light. **b, c** Optical reflection images obtained with linearly polarized illumination. The relative orientation of the gyroid domains (*red inserts*) and the light polarization (*the white arrows*) produces a strong birefringence effect. **d** Variation of the reflection spectra as a function of the angle between the in-plane [110] gyroid direction and the incident light polarization

As expected the transmission decayed exponentially with thickness. The decaying rates of the exponential fittings were then calculated for different wavelengths and plotted in Fig. 4.9. Interestingly the decay length were longer for wavelengths close to the strongest propagating modes at around 550 nm.

4.4.3 Disorder Effects

The contribution of the periodicity of the gyroid structure was investigated by inducing a structural distortion via atomic diffusion [9] at high temperatures. Atomic diffusion in nanostructures often leads to surface area minimization. The constant mean curvature of the gyroid surface suggests high stability of the gold gyroid against atomic diffusion, but nonetheless, atomic diffusion is favoured to happen towards the grains with more favourable crystal planes on the surface [10].

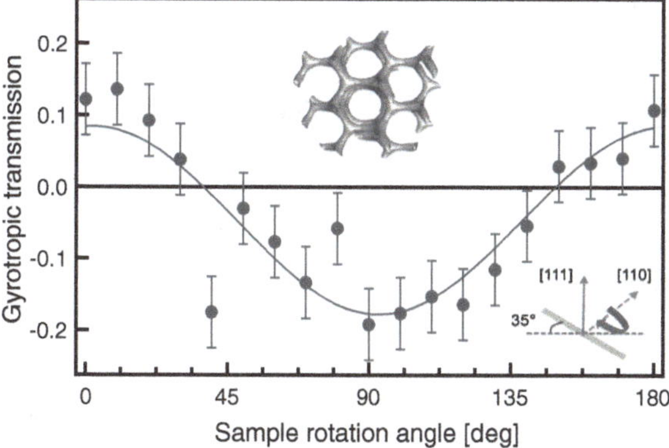

Fig. 4.7 Gyrotropic behaviour of the gold gyroid with a 50 nm unit cell across different directions. The sample with the [110] out-of-plane orientation was inclined by 35° and rotated around the [110] surface normal to access the chiral [111] direction. The gyrotropic transmission was calculated as the difference between the left and right circular polarizations averaged over frequency region between 600 and 750 nm. The line is a guide for the eye. Courtesy of Vignolini et al. [8]

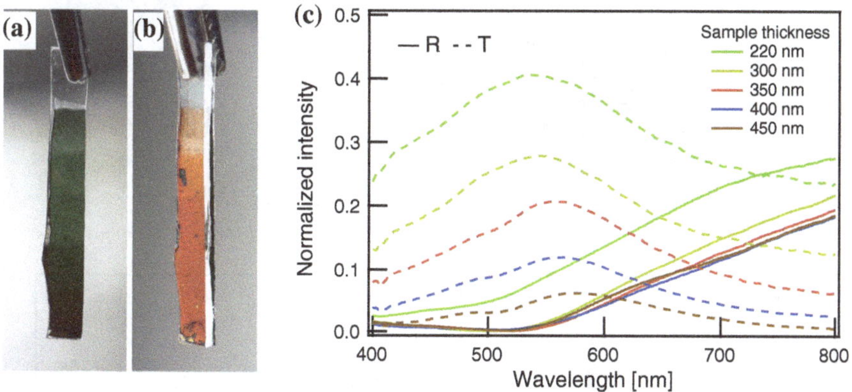

Fig. 4.8 **a, b** A gyroid film with increasing thickness photographed in transmission and reflection. **c** Reflection and transmission spectra for increasing thicknesses

The result was a distortion of the gyroid structure by the assembly of the gyroid strut into thicker clusters. This process was activated heating the sample at 100 °C and was followed for 12 h during which reflection and transmission spectra were taken at constant time intervals (Fig. 4.10). While small deformations of the gyroid structure did not significantly affect the effective electron mass and the reflectivity, they did modify the propagation modes. Interestingly, the transmission spectrum was not only reduced in intensity but drastically modified in its shape, suggesting the contribution of different modes of light propagation through the gyroid network.

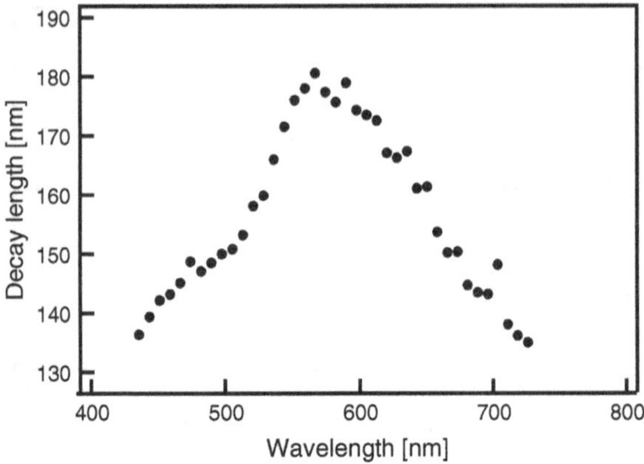

Fig. 4.9 Transmission decay length extracted from the exponential decays with thickness for different wavelengths

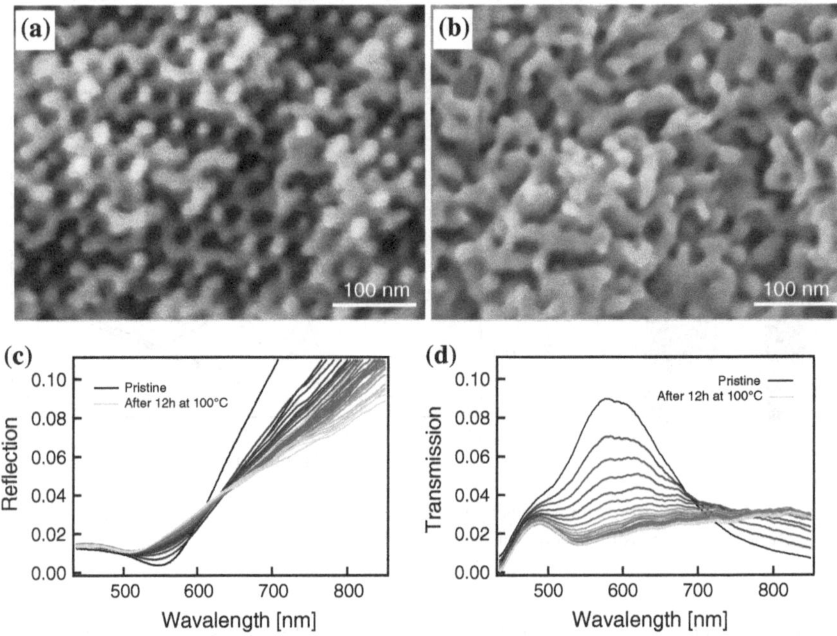

Fig. 4.10 SEM image of the gold gyroid before (**a**) and after (**b**) heating to 100 °C for 12 h. The atomic diffusion activated at high temperature produced a structural distortion that was investigated in reflection (**c**) and transmission (**d**). Scan intervals: 20 min

More drastic distortions were produced at 200 °C (Fig. 4.11). At this temperature the gold gyroid struts were modified into thicker clusters that responded optically very similarly to bulk gold.

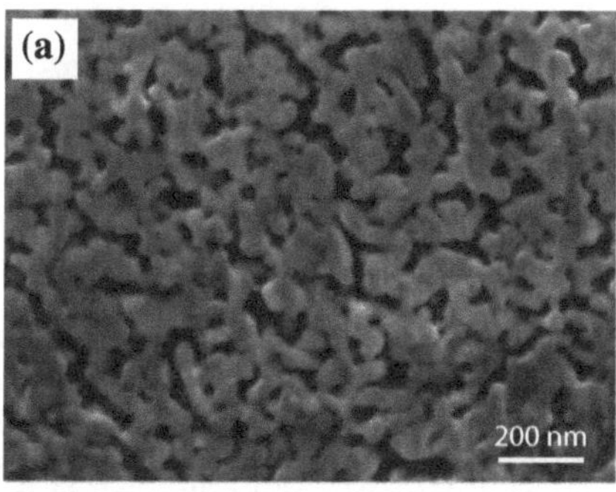

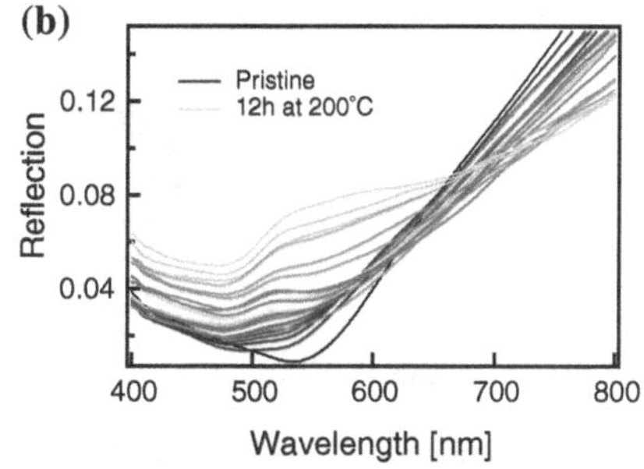

Fig. 4.11 **a** SEM image of the gold gyroid after heating to 200 °C for 12 h. **b** Reflection spectra for scan intervals of 20 min

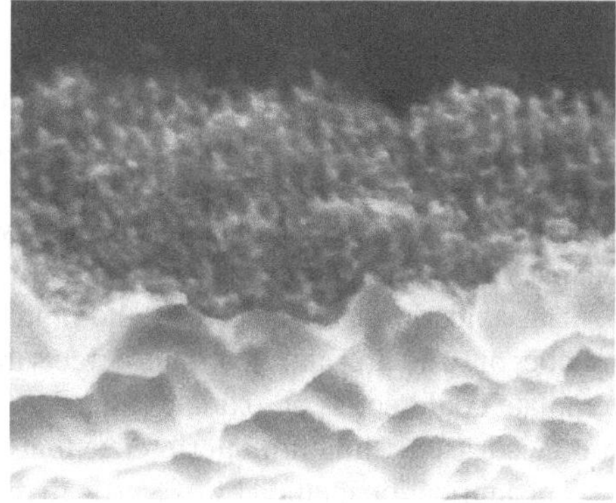

Fig. 4.12 SEM image showing a free standing gold gyroid film viewed from the *bottom*. The continuous thin layer of gold is formed during the electrodeposition at locations where the block copolymer film had detached from the FTO substrate. The morphology of this film resulted from the substrate FTO roughness

4.4.3.1 Experimental Variations

The structural deformation activated by atomic diffusion can also explain the small variations of the reflectivity found for different samples. During the removal of the polymer template via plasma etching, the samples were exposed to high temperatures (between 50 and 100 °C) for 10 min, inducing small structural distortions.

Experimental variations were also found in transmission. This was strongly affected by the formation of a continuous thin layer between the substrate and the gyroid. Local detachments from the block copolymer film to the FTO substrate were filled with gold during the electrodeposition, producing a continuous thin layer that reduced the transmission (Fig. 4.12). The formation of this layer was partially reduced by the substrate functionalization discussed in the previous chapter.

Fig. 4.13 Oxidation effects to the silver gyroid after 6 h (*left*) and 1 h (*right*) after polymer template removal. The color change is a combination of the plasmonic properties of the silver gyroid and the dielectric silver oxide. The *right* sample was patterned by a micromanipulator producing the *square* array

4.4.4 Double and Silver Gyroids

Due to the lower optical losses, silver is often considered the most promising metal for a metamaterial acting at optical frequencies [11]. Nonetheless, its practical application in nanostructures is severely limited by the strong tendency to oxidize. Silver gyroid indeed oxidized soon after the removal of the polymer template, producing a change in coloration that varied with the time and across the samples (Fig. 4.13).

The oxidation is enhanced by the high surface area and promoted by the high number of structural and electronic defects in nanostructured silver [12]. The polymer template could act as an oxidation barrier but the thin unfilled part of the template produced strong thin film interference that impeded the optical characterization.

The fabrication method by overgrowth removal, discussed in the previous chapter, enabled the creation of silver gyroid films with flat surfaces that were perfectly filled by the block copolymer template. The polymer, therefore, acted as an oxidation barrier without producing the contribution of additional optical layers.

Silver and gold gyroids were compared for similar conditions, i.e. with the polymer template still in place and perfectly filled by the metal. As discussed in the next chapter, the presence of the polymer as surrounding medium produces a further reduction of the plasma frequency and a red shift of the reflection spectra. It is reasonable to assume that the effect of the polymer produces similar optical effects in gold and silver gyroids.

Both the reflection and transmission of silver gyroids were characterized by a shift to shorter wavelengths of about 100 nm with respect to the gold gyroid (Fig. 4.14), in agreement with the lower plasma frequency of bulk silver. It is important to stress that these silver and gold gyroid films were produced from similar polymer samples, with a very precise control of the layer thickness.

The effect of the second gyroid network in the double gyroid morphology was also investigated for both gold and silver. As previously described, the double gyroids were produced using diblock copolymer and consisted of a non-chiral structure of two interwoven separate networks.

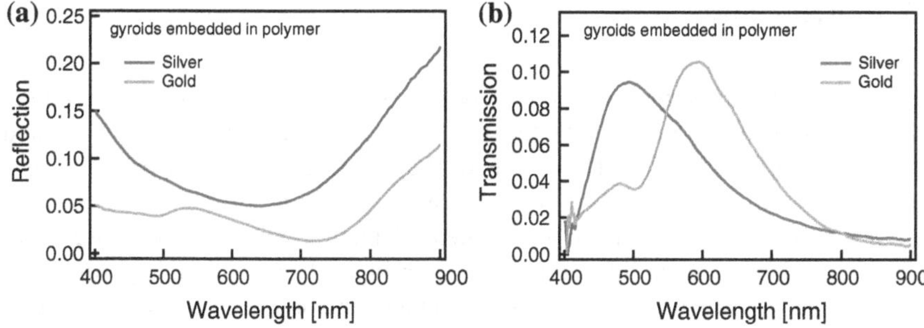

Fig. 4.14 **a** Reflection and **b** transmission spectra of gold and silver single gyroids. The samples were characterized with the block copolymer template in place to prevent silver oxidation. As discussed in the next chapter, the surrounding medium with a refractive index of $n = 1.6$ (e.g. for polystyrene) produces a red-shift of the reflection spectra. The reflection spectra were for a measured on film sample 1 μm thick sample, while the transmission spectra were taken for a 450 nm thick sample

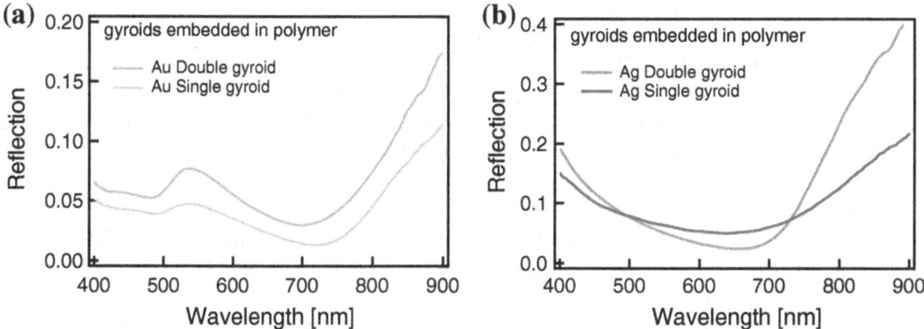

Fig. 4.15 **a** Reflection spectra of single gyroid made in gold and silver. **b** Reflection spectra of double gyroid made in gold and silver

The second gyroid network was predicted [13] to produce a capacitive effect and act as metal/insulator/metal waveguide, leading to a different light propagation mechanism that stems from coupled surface plasmon resonances of closed loops of the gyroid interwoven networks.

The optical response of the single versus double gold gyroids were extremely similar, despite of the random orientation of the double gyroid domains and slightly different filling fractions ($f = 38\,\%$ in the double gyroid and $f = 30\,\%$ in the single gyroid). However, the difference between single and double gyroids was more evident in the silver gyroids (Fig. 4.15).

4.5 Conclusions

These gyroid-based structures effectively behaved like metamaterials at visible wavelengths. The two main characteristics were a reduced plasma frequency and enhanced light transmission. The effects of the film thickness and structural ordering were discussed, showing the existence of plasmonic propagating modes in the gyroid structure. This metamaterial exhibited orientation-dependent response under linearly polarized incident light and optical chirality. Silver and double gyroid were also discussed. However, the comparison of single and double gyroids was difficult due the different out-of-plane orientation.

References

1. Williams DB, Carter CB (1996) The transmission electron microscope. Springer, New York
2. Fuchs K (1938) The conductivity of thin metallic films according to the electron theory of metals. Math Proc Cambridge Philos Soc 34:100–108
3. Durkan C, Welland ME (2000) Size effects in the electrical resistivity of polycrystalline nanowires. Phys Rev B 61(20):14215
4. Sondheimer EH (1952) The mean free path of electrons in metals. Adv Phys 1(1):1–42
5. Pendry JB, Holden AJ, Robbins DJ, Stewart WJ (1998) Low frequency plasmons in thin-wire structures. J Phys Condens Matter 10(22):4785–4809
6. Oh SS, Demetriadou A, Wuestner S, Hess O (2012) On the origin of chirality in nanoplasmonic gyroid metamaterials. Adv Mater. doi:10.1002/adma.201202788
7. Oskooi AF, Roundy D, Ibanescu M, Bermel P, Joannopoulos JD, Johnson SG (2010) Meep: a flexible free-software package for electromagnetic simulations by the FDTD method. Comput Phys Commun 181(3):687–702
8. Vignolini S, Yufa NA, Cunha PS, Guldin S, Rushkin I, Stefik M, Hur K, Wiesner U, Baumberg JJ, Steiner U (2012) A 3D optical metamaterial made by self-assembly. Adv Mater 24(10):OP23–OP27 (Deerfield Beach, Fla.)
9. Kuipers L, Hoogeman MS, Frenken JWM (1993) Step dynamics on Au (110) studied with a high-temperature, high-speed scanning tunneling microscope. Phys Rev Lett 71(21):3517
10. Ehrlich G (1991) Direct observations of the surface diffusion of atoms and clusters. Surf Sci 246(1):1–12
11. Tassin P, Koschny T, Soukoulis CM (2012) Effective material parameter retrieval for thin sheets: theory and application to graphene, thin silver films, and single-layer metamaterials. Phys B 407:4062–4065
12. Franke ME, Koplin TJ, Simon U (2006) Metal and metal oxide nanoparticles in chemiresistors: does the nanoscale matter? Small 2(1):36–50
13. Hur K, Francescato Y, Giannini V, Maier SA, Hennig RG, Wiesner U (2011) Three-dimensionally isotropic negative refractive index materials from block copolymer self-assembled chiral gyroid networks. Angew Chem Int Ed Engl 50(50):11985–11989

Chapter 5
Tuning Methods

5.1 Introduction

The optical response of the gold gyroid metamaterial was tuned using three different methods as summarized in Fig. 5.1. In the first, the structural dimensions of the mesoscopic unit cell were controlled by varying the molecular weight of the structure-forming block copolymer. In the two other strategies using post-processing, the tuning was achieved by controlling the metal filling fraction of the plasmonic structure and by filling different refractive index media into the metal scaffold [1].

The experimental results showed good agreement with finite difference time domain (FDTD) [2] calculations of the full 3D structured and also matched an approach based on a three-helix model (THM) [3]. The FDTD calculations and THM model were provided by Sang Soon Oh and Angela Demetriadou.[1]

The THM described the gyroid architecture more accurately than the wire array model discussed in the previous chapter and assumed that the gyroid's internal structure derives from a network of interconnected metallic helical wires. In particular, there are two types of helices, visible in the [100] direction, with different radii [4]. The THM showed that the optical properties close to the cut-off wavelength are determined by the smaller helices (see Fig. A.1 in the Appendix for a 2D representation of the structure). This cut-off wavelength originates from the continuous nature of the metallic helices and is related to the plasma wavelength. No wave can propagate in the gyroid for wavelengths larger than the plasma wavelength leading to high reflection at these wavelengths. Conversely, three propagating modes were identified just below the plasma wavelength, highly confined within the smaller helices.

As a result the THM model allowed to calculate the plasma wavelength from the following equation which makes use of the geometric parameters of the smaller helices:

---

[1] Ortwin Hess' group, Department of Physics, Imperial College London.

© Springer International Publishing Switzerland 2015
S. Salvatore, *Optical Metamaterials by Block Copolymer Self-Assembly*, Springer Theses,
DOI 10.1007/978-3-319-05332-5_5

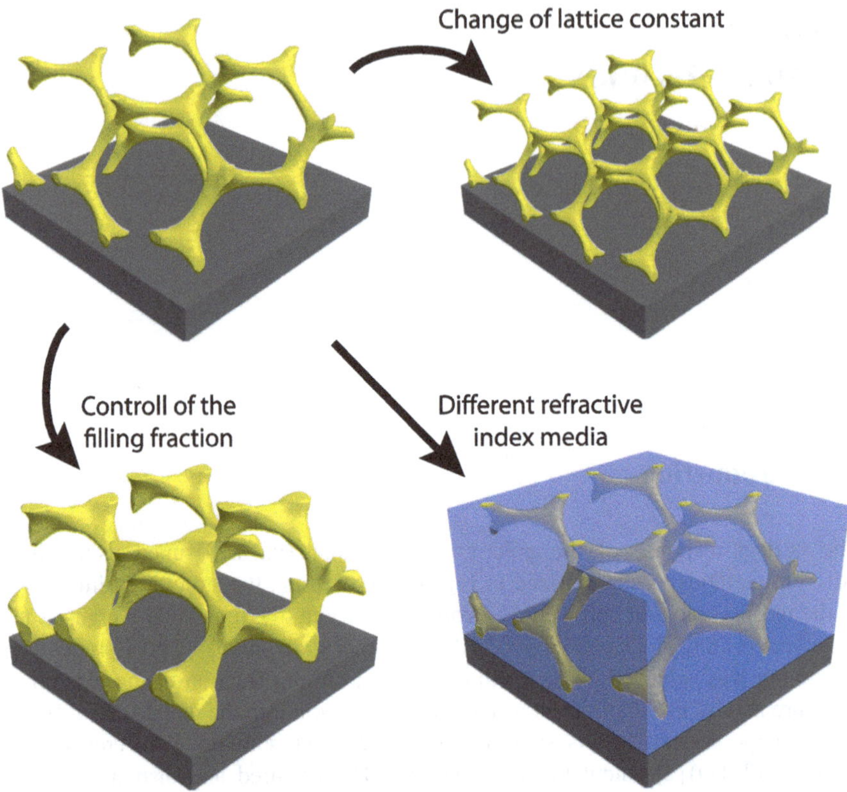

Fig. 5.1 Schematic representation of the tuning mechanisms described in this chapter: the optical properties were tuned by structural and environmental variations controlling the lattice constant, the filling fraction and the refractive index of the surrounding medium

$$\lambda_{\mathrm{p}} \approx c_1 a \left\{ \frac{1}{\epsilon_{\mathrm{Au}} f c_2} + \frac{c_3}{\epsilon_n \left[c_4 \sqrt{f} - \sqrt{c_5 + c_6 \ln \left(c_7 / \sqrt{f} \right)} \right]^2} \right\}^{-\frac{1}{2}}, \tag{5.1}$$

where f and a are the filling fraction and unit cell of the gyroid, ϵ_n is the dielectric constant of the medium (with refractive index n) the gyroid is immersed in, ϵ_{Au} is the dispersive permittivity of gold, and c_{1-7} are geometrical coefficients (see Appendix A.1).

The optical properties can therefore be tailored by varying the geometrical parameters of the metallic gyroid and the dielectric material it is filled with.

The results presented in this chapter were reviewed and published on Advanced Materials [1].

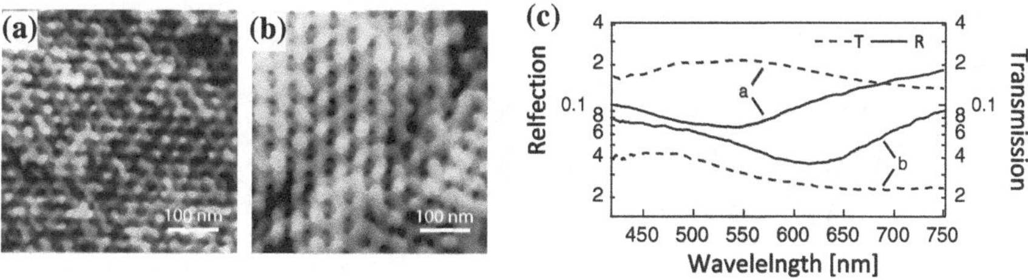

Fig. 5.2 **a, b** SEM images of gold gyroid metamaterials obtained from ISO terpolymers of two different molecular weights, yielding lattice constants of 35 and 50 nm, respectively. **c** Corresponding transmission and reflection spectra for unpolarized incident light. Film thickness ~300 nm

5.2 Unit Cell Control

The influence of the lattice constant on the optical properties was studied by comparing two gyroid morphologies obtained from ISO block copolymer with two different molecular weights but identical block volume fractions. The used ISO molecular weights of 33 and 53 kg·mol^{-1} correspond to measured unit cell sizes of 35 and 50 nm, respectively, while maintaining the same filling fraction of 30 %. The scanning electron microscopy (SEM) images in Fig. 5.2a, b show the gold gyroid fabricated from the two ISO triblock copolymers, Fig. 5.2c compares their reflection and transmission spectra.

As expected, for both lattice sizes the highest reflection values were observed on the long wavelength side of the spectrum and a dip in the spectrum was found when wave propagation in the gyroid slab is permitted (i.e., for $\lambda < \lambda_p$). Furthermore, as predicted by the THM model, the $a = 50$ nm unit cell has a lower plasma frequency compared to the gyroid with $a = 35$ nm.

The reflectivity of the smaller gyroid had a minimum at a wavelength of ~550 nm, which shifted to ~620 nm for the larger gyroid. For larger unit cells and identical filling fractions f (i.e., identical effective electron densities) the gyroid struts were thicker and, as a result, the induced self-inductance was stronger, corresponding to higher effective electron mass per unit cell. Since the effective electron mass is linearly proportional to the plasma wavelength, larger lattice constants led to an increase in the plasma wavelength, as observed in Fig. 5.2c. This property is also clearly evident in Eq. 5.1, where for constant f, λ_p is linearly dependent on a. A bigger value of a however significantly increased absorption, resulting in a significantly lower transmission for the larger gyroids.

5.3 Filling Fraction Variation

The manufactured gyroids could be post-processed by further gold electrodeposition. In the absence of the surrounding polymer matrix, the entire network structure acted as a cathode and gold growth occurred homogeneously to thicken the gyroid struts

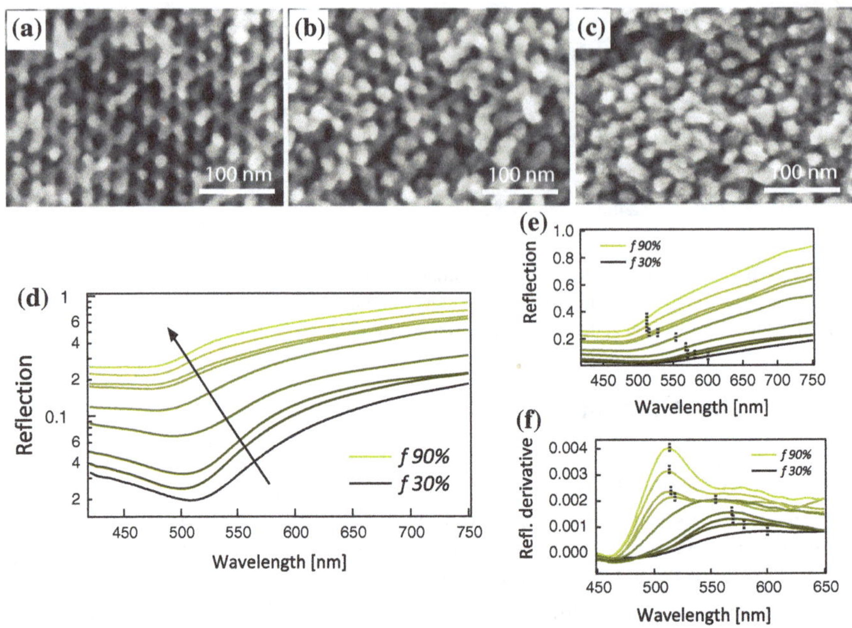

Fig. 5.3 **a–c** SEM images of gold gyroids with $a = 35$ nm and filling fractions of 30, 60, and 75 %, respectively. **d**, **e** Reflection spectra for filling fractions varying from 30 to 90 % in ~8 % increments, in logarithmic and linear scales. **f** First derivative of the reflection spectra highlighting the inflection points (*dashed vertical lines*). The point of inflection in the reflection spectra was measured calculating the maximum value of the first derivative

without affecting the symmetry of the network. This process continuously increased the filling fraction by thickening the struts, leaving all other parameters unchanged.

Figure 5.3 shows the effect of the filling fraction increase from 30 to 90 %, resulting in a reduction in the plasma wavelength and thereby causing a blue shift of the reflectivity edge to shorter wavelengths. As the filling fraction approached 100 % (i.e. solid gold) the reflection spectra indeed approached that of the gold and the reflection minimum disappeared (Fig. 5.3d).

As the plasma wavelength defined in Eq. 5.1 could not easily be extracted from reflection and transmission spectra, it was convenient to introduce the plasma edge wavelength λ_{pe}, which is the wavelength at the point of inflection of the reflectivity (i.e. the wavelength where the first derivative of the spectrum has a maximum, which was easily identified in experimental and calculated spectra, Fig. 5.3e, f. Although the plasma edge λ_{pe} and plasma wavelength λ_p are not identical, they have the same behaviour and differ only by small shift as the gyroid metamaterial is tuned.

A quantitative analysis of the results in Fig. 5.3d is shown in Fig. 5.4, where the plasma edge wavelength is plotted versus the filling fraction. This analysis reveals a continuous shift in λ_{pe} of more than 100 nm between solid gold and a gold gyroid with filling fraction of 30 %. These results were in excellent agreement with the results

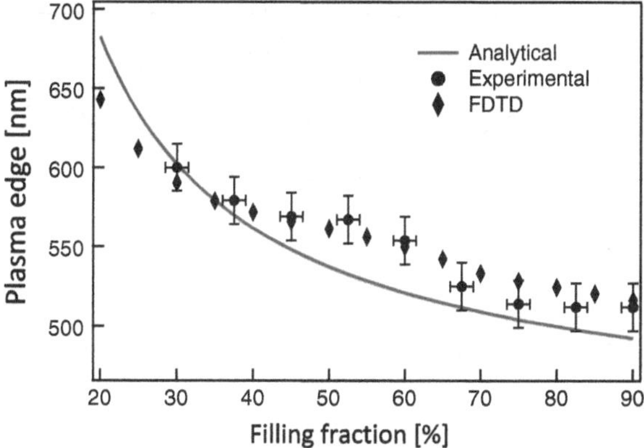

Fig. 5.4 Variation of the plasma edge wavelength λ_{pe} with filling fraction. Experimental data (*circles*) compared to the THM analytical model (*line*) and a FDTD calculation (*diamonds*)

of FDTD calculations, as well as with the analytical THM model for the plasma wavelength.

For $f \rightarrow 100\,\%$ the plasma wavelength calculated from the THM does not asymptotically lead to the plasma wavelength of bulk gold at 165 nm. Although the THM is unable to predict the behaviour of the gold gyroid at very high filling fractions it is also realistic to expect a drastic change in the plasma frequency from bulk gold to gyroid with $f \rightarrow 100\,\%$ due to the continuous porous network still present at very high filling fractions. Therefore it is sensible to assume the model valid up to 90 % of filling fraction.

Interestingly, despite of the increase of metal content, the filling fraction increment generates a reduction in the optical absorption, as shown in Fig. 5.5. While the transmission slowly decays with f, the reflection is rapidly enhanced, producing higher absorption for lower filling fraction gyroids than from bulk gold.

The change of filling fraction also affected the linear dichroism in the [110] direction that has been discussed in the previous chapter. Figure 5.6 shows the reflection spectra for different filling fractions and with incident polarizations perpendicular and parallel to the in-plane [100] direction of the gyroid lattice. For ~50 % filling fraction, the linear dichroism is lost. This can be explained considering the origin of the linear dichroism, that arises from a variation of the coupling of polarized light with the differing gyroid lattice symmetries: as the anisotropy of the gyroid was reduced with increasing filling fraction, the dichroism disappeared.

5.4 Surrounding Medium Effects

A further way to reversibly tune the optical response of the gyroid metamaterial involved changing the dielectric medium infiltrated into the gyroid. Figure 5.7 shows reflection spectra of the 35 nm unit cell gyroid with a fill fraction of 30 % immersed

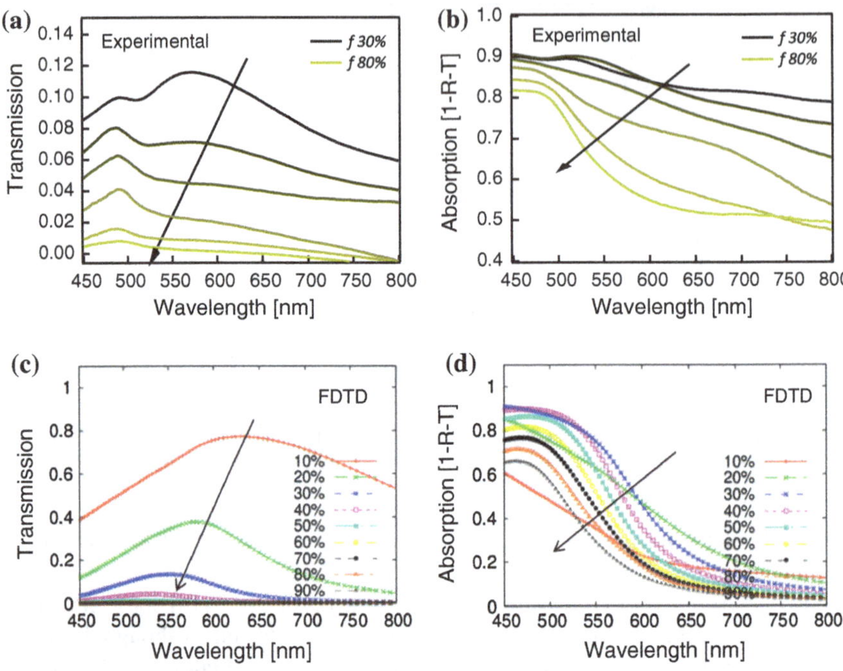

Fig. 5.5 **a–d** Transmission and absorption spectra from experimental measurements and FDTD calculations

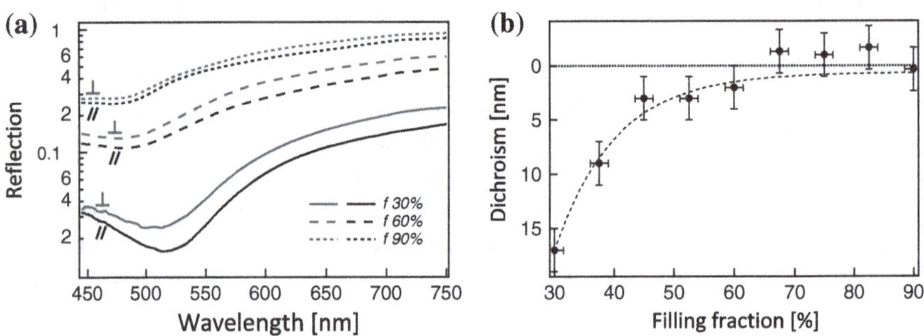

Fig. 5.6 Linear dichroism of the gold gyroid in the [110] direction as a function of the filling fraction. **a** Reflection spectra with incident polarization perpendicular (*red*) and parallel (*black*) to the in plane [100] direction of the gyroid lattice. **b** Experimental data for the linear dichroism as function of the filling fraction. The dichroism measure was calculated as the difference of the plasma edge wavelengths of perpendicular and parallel polarizations to the [100] direction. The *dotted line* is a guide to the eye

in media with increasing refractive index. The plasma wavelength increased linearly with n as illustrated by the variation of the reflectivity curves in Fig. 5.7b. This variation agreed quantitatively with the numerical and analytical results.

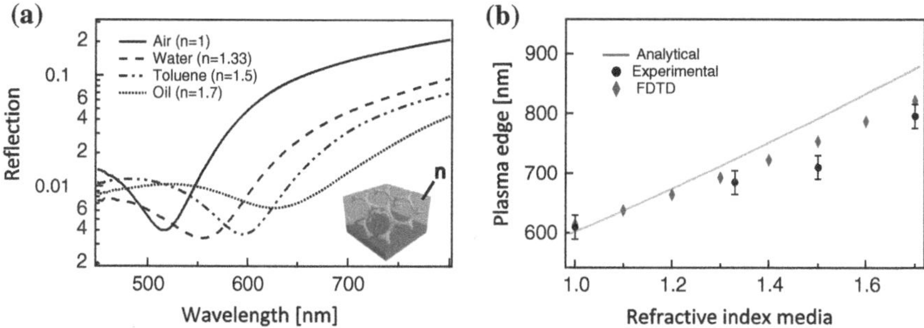

Fig. 5.7 **a** Reflection spectra for unpolarized light incident on the $a = 35\,\text{nm}$ gyroid filled with dielectric media: air ($n = 1$), water ($n = 1.33$), toluene ($n = 1.5$), and Cargille Refractive Index Liquid Series B ($n = 1.7$). **b** Variation of the plasma edge wavelength λ_{pe} for different refractive index media compared to the THM analytical model and FDTD calculations

5.5 Conclusions

In conclusion, three successful approaches were demonstrated to tune the optical behavior of gyroid self-assembled optical metamaterials. These methods can also be applied to other 3D metamaterials.

Optimised structural parameters led to a transmission through 300 nm thick layers of gyroid-structured gold of up to 30 %. Increasing the lattice dimensions in the gyroid metamaterial reduced the plasma frequency and the transmittivity, which was higher for the small unit cell with lower filling fraction. Increasing the gold strut thickness (i.e. the filling fraction) caused a continuous deformation in the shape of the reflectivity curves and increased the plasma frequency towards solid gold. Finally, the optical characteristics could be reversibly tuned by adjusting the refractive index of the medium that is infiltrated into the gyroid pore structure, revealing a linear variation of the plasma edge with n.

References

1. Salvatore S, Demetriadou A, Vignolini S, Oh SS, Wuestner S, Yufa NA, Stefik M, Wiesner U, Baumberg JJ, Hess O, Steiner U (2013) Tunable 3D extended self-assembled gold metamaterials with enhanced light transmission. Adv Mater (Deerfield Beach, Fla.) 25(19):2713–2716
2. Oskooi AF, Roundy D, Ibanescu M, Bermel P, Joannopoulos JD, Johnson SG (2010) Meep: a flexible free-software package for electromagnetic simulations by the FDTD method. Comput Phys Commun 181(3):687–702
3. Demetriadou A, Oh SS, Wuestner S, Hess O (2012) A tri-helical model for gyroid metamaterials. New J Phys 14. doi:10.1088/1367-2630/14/8/083032
4. Oh SS, Demetriadou A, Wuestner S, Hess O (2012) On the origin of chirality in nanoplasmonic gyroid metamaterials. Adv Mater. doi:10.1002/adma.201202788

Chapter 6
Hollow Gyroid

6.1 Introduction

The gyroid geometry topology discussed above was used as a starting point to create a further complex architecture, the hollow-gyroid with an increased surface area and strongly enhanced optical transmission.

The chapter will commence describing the fabrication technique and the optical properties of the gold hollow gyroid, compared to the gold gyroid discussed in the previous chapters, referred to 'normal gyroid'. I will then discuss the properties of 'composite' metamaterials employing different metals and dielectrics in the gyroidal structure.

A similar fabrication methodology was also pursued to fabricate amorphous carbon hollow gyroids which are relevant in high performance battery applications.

Finally, I will show a similar fabrication methodology to fabricate the inverse gyroid structure mimicking the majority phase of the self-assembled double gyroid block copolymers.

To allow the comparison with the previous results, the characterizations presented in this chapter were all done using 300 nm thick samples.

6.2 Hollow Gyroid Fabrication

The ISO triblock copolymer was spin coated onto a FTO-glass substrate, thermally annealed and exposed to UV light to remove the minor polyisoprene phase, as described in Chap. 3. The polymer template was then back-filled with nickel and removed by plasma etching.

The nickel gyroid then acted as working electrode to homogeneously grow gold around the nickel struts. Finally, the inner nickel core was etched away by immersion in iron chloride ($FeCl_3$) for 30 min, followed by rinsing with deionised water. The fabrication steps, the respective SEM images and the energy dispersive X-ray spectroscopy (EDX) are shown in Fig. 6.1.

© Springer International Publishing Switzerland 2015

S. Salvatore, *Optical Metamaterials by Block Copolymer Self-Assembly*, Springer Theses,
DOI 10.1007/978-3-319-05332-5_6

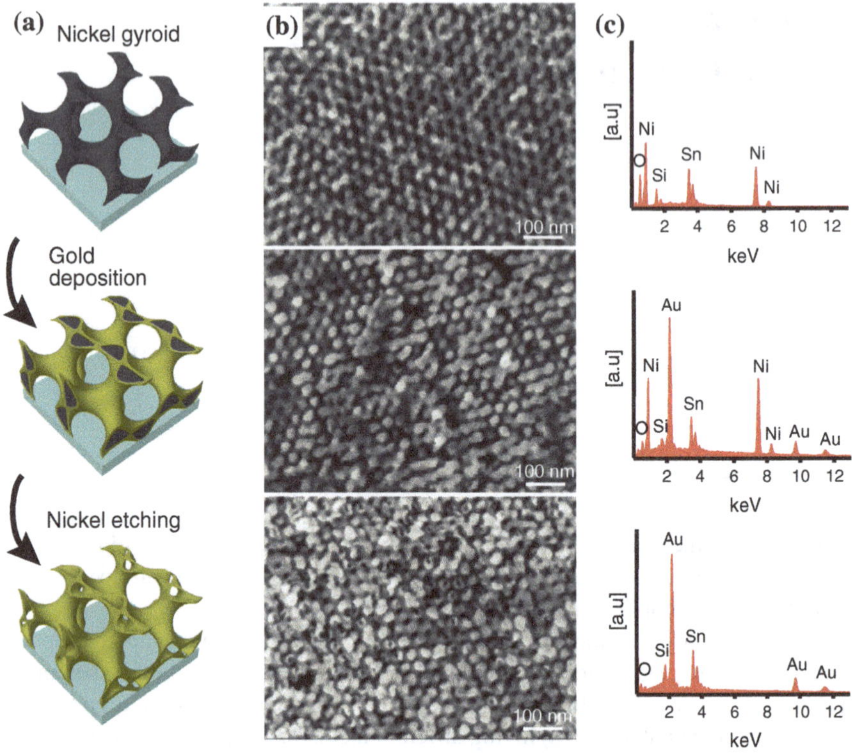

Fig. 6.1 **a** Schematic representation of the hollow gyroid fabrication process. A nickel gyroid is coated with a thin layer of gold by electrodeposition and the nickel core is selectively etched away by $FeCl_3$. **b** SEM images of the of the three fabrication steps: after the gold coating the gyroid struts appear thicker and after the nickel etching the hollow cavities are clearly visible along a strut cross-section. **c** Energy dispersive X-ray spectroscopy (EDX) measurements for the three fabrication steps. The ratio between nickel and gold masses at the second steps allows to identify the filling fraction of gold. The complete nickel removal is confirmed by the EDX analysis after $FeCl_3$ etching. The Tin was detected from the FTO substrate

Interestingly, despite of the gold coating, the nickel core was readily removed by $FeCl_3$. It is reasonable to assume that local defects in the gold coating served as $FeCl_3$ access channels to the Ni-core. Moreover, the thin layer of gold coating was less than 5 nm thick and consisted of grains with similar dimensions. Interstitial diffusion at the grain boundaries might have also contributed to the etching process.

The electrodeposition of gold onto the nickel gyroid struts was obtained by applying a low steady potential of (−0.7 V) for times ranging from 10 to 40 s, depending on the desired final gold filling fraction.

The gold filling fraction was measured by energy dispersive X-ray spectroscopy (EDX) [1] before the nickel core removal. EDX is an analytical technique used for the elemental analysis that allows a quantitative measurement of the elemental composition. As the nickel fill fraction was known from the initial volume fraction of the polyisoprene phase (~30 %), the elemental ratio between gold and nickel determines the gold filling fraction. Figure 6.1 shows the 1:1 ratio that was used to compare the optical properties of the hollow gyroid with the normal gyroid. The complete removal of the nickel core was also examined by EDX, as shown in Fig. 6.1.

6.2.1 Optical Properties

6.2.1.1 Comparison of Hollow and Normal Gyroids

Reflection and transmission spectra of a normal gyroid were compared with a hollow gyroid with equal filling fraction and are shown in Fig. 6.2. Despite incorporating the same amount of gold, the reflection and transmission intensities were considerably higher in the hollow gyroid. The reflection edge of the hollow gyroid is shifted to smaller wavelengths with respect to the normal gyroid. This shift is similar to that produced by increasing the filling fraction of the normal gyroid, seen in the previous chapter. This behaviour is reasonable considering that in both the hollow gyroid and the high-f normal gyroid have increased outer strut radii.

Remarkably, the transmission of the hollow gyroid was enhanced by a factor of 2 with respect to the normal gyroid. As previously discussed, the high transmission through the gold gyroid was caused by the propagating modes through the gyroid architecture. In the hollow gyroid, the surface propagating modes can propagate along both the inner and outer interfaces of the hollow struts, increasing the transmission. Moreover, the pipe-like morphology have thinner walls compared to the strut-width of a normal gyroid with an equal filling fraction. Consequently, the evanescent modes which decay exponentially in the metal, experience lower losses in the hollow structure, as seen in Fig. 6.2 where the sum of transmission and reflection is substantially increased compared to the normal gyroid.

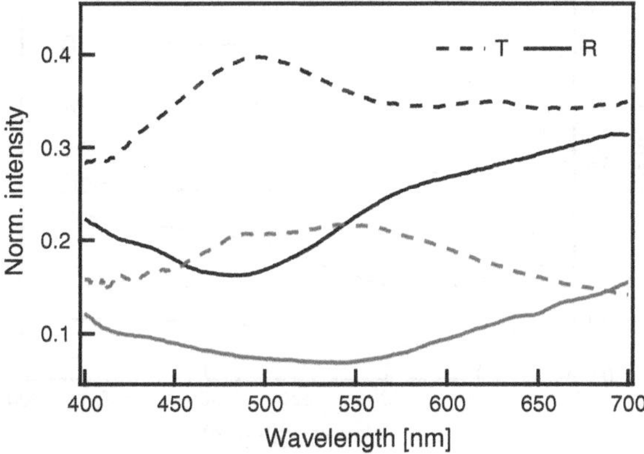

Fig. 6.2 Reflection and transmission spectra of the gold hollow gyroid compared with the normal gyroid discussed in the previous chapter. Film thickness 300 nm

6.2.1.2 Thin Hollow Gyroid

Hollow gyroid with thinner strut wall widths were produced by electrodepositing gold for shorter times. The gold filling fraction was again measured by EDX, comparing the amount of gold to the known fraction of nickel. Reducing the gold fill fraction from 30 to 20 and 10 % increased the transmission to ~48 and ~58 %, which are remarkable values for 300 nm thick gold films (Fig. 6.3).

To compare the transmission of different filling fractions the transmission peaks T can be normalized to the air fraction. The so-normalized transmittance increases from $T_{\mathrm{norm}} = 57\,\%$ for $f = 30\,\%$ to $T_{\mathrm{norm}} = 60\,\%$ and $T_{\mathrm{norm}} = 64\,\%$ for $f = 20\,\%$ and $f = 10\,\%$ respectively. Where T_{norm} was calculated multiplying the transmission peak by the filling fraction.

6.2.1.3 Composite Hollow Gyroid

The gold coated nickel gyroid is a metal-composite optical metamaterial. Although the nickel does not have plasmonic properties at visible wavelengths, there has been an increasing interest in nanoantennas made of a ferromagnetic material due to their ability to combine an optical plasmon resonance with strong magnetic properties [2, 3].

Optical antennas are devices designed to convert optical radiation into localized energy, enhancing the interaction between light and matter [4]. However, the plasmonic behaviour in ferromagnetic materials such as nickel is strongly reduced by losses compared to noble metals such as gold. A common strategy to overcome this strong damping is the development of hybrid structures consisting of noble and ferromagnetic materials to increase the plasmon response [5, 6]. Therefore, the study of the combined optical properties of composite Ni-Au gyroids is of interest.

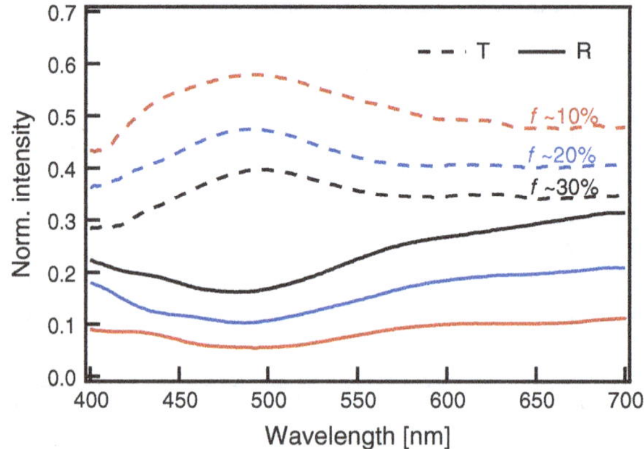

Fig. 6.3 Reflection and transmission spectra of gold hollow gyroids for filling fractions of 30, 20 and 10 %. The filling fraction was measured by EDX analysis before nickel core removal

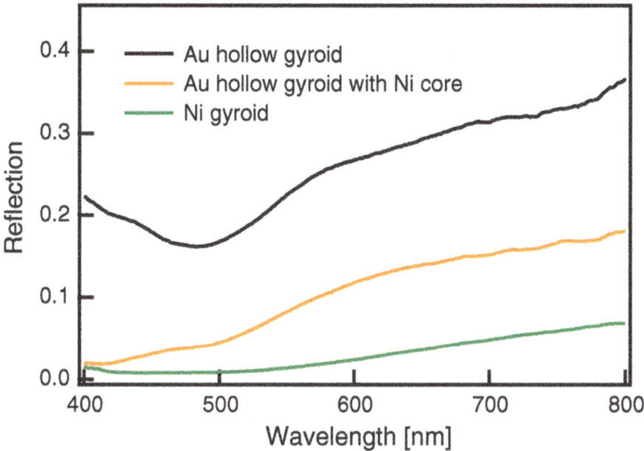

Fig. 6.4 Reflection spectra of the gold hollow gyroid compared to the gyroid with a nickel core and to the nickel-only gyroid. The transmittance of the nickel-containing structures was negligible and are not shown

As expected, the nickel gyroids had strong optical losses and high light absorbtion in the entire visible spectra as shown in Fig. 6.4. Interestingly, the gold gyroid with nickel core displayed an optical response similar to that of the gold hollow gyroid, suggesting the importance of plasmonic modes on the outer gold shell despite of the high losses in the nickel core.

6.2.1.4 Insulator-Metal-Insulator Metamaterial

Hollow gyroids with dielectric cores were fabricated by electrodepositing gold on nickel oxide (NiO) cores. Nickel gyroids were heated at 500 °C in air for 12 h to promote complete oxidation. The thin layer of gold was then deposited using a pulsed voltage of -1.1 V for 10 cycles of 0.5 s. The higher deposition voltage was required because of the low NiO conductivity.

The NiO effectively acted as a dielectric with a refractive index $n = 2.18$, forming an asymmetric insulator-metal-insulator metamaterial. When two metal-dielectric interfaces are placed close together, the surface plasmons of each interface overlap, giving rise to coupled modes. Cylindrical waveguides with high refractive index core and metal cladding have recently been investigated [7], showing that the coupled modes of asymmetric interfaces (metal-dielectric$_1$ and metal-dielectric$_2$) support slow-light propagation, producing degenerate forward and backward waves.

In the hollow gyroid with asymmetric dielectrics, the reflection spectra had two distinctive edges as shown in Fig. 6.5. In the previous chapter the effect of an increased refractive index of the medium surrounding the gyroid struts reduced the plasma frequency and produced a red-shift of the reflection spectrum. The two reflection edges in the asymmetric hollow gyroid can therefore be explained as a result of the two interfaces, the metal-NiO interface producing the shift at ~700 nm and the metal-air

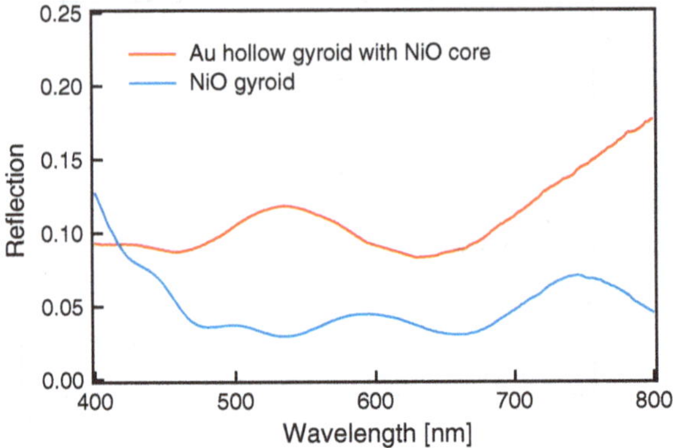

Fig. 6.5 Reflection spectra of the gold gyroid with a nickel oxide (NiO) core. The refractive index of NiO is $n = 2.18$

interface producing the edge at shorter wavelengths. In Fig. 6.5 the hollow gyroid with asymmetric dielectrics is shown together with the reflection spectrum of a NiO-only gyroid, showing the characteristic fringes of dielectric thin film interference.

6.3 Carbon Hollow Gyroid

Similarly to the fabrication process of the gold hollow gyroid, the nickel gyroid structure was employed as sacrificial frame to fabricate amorphous carbon hollow gyroids. Nickel gyroid samples were heated to 500 °C for times varying from 5 to 30 min in a custom-build cold wall reactor and exposed to C_2H_2 to initiate graphitic carbon growth by chemical vapour deposition (CVD). This process was carried out by Piran Kidanmbi[1] [8]. The nickel core was then removed by $FeCl_3$ and the samples were characterised by SEM (Fig. 6.6).

The hollow carbon gyroid structure is interesting as high surface are anode in lithium ion batteries. Carbon materials are used in battery systems because of to their unique combination of chemical, electrical, mechanical, and thermal properties [9]. The performance of lithium ion batteries, such as the charge/discharge capacity and voltage profile, depend strongly on the microstructure of both electrodes [10]. The high surface-to-volume ratio of the carbon hollow gyroid makes this structure ideal to design of carbon/lithium ion batteries with high charge/discharge capacity.

[1] The Hofman group, Department of Engineering, University of Cambridge.

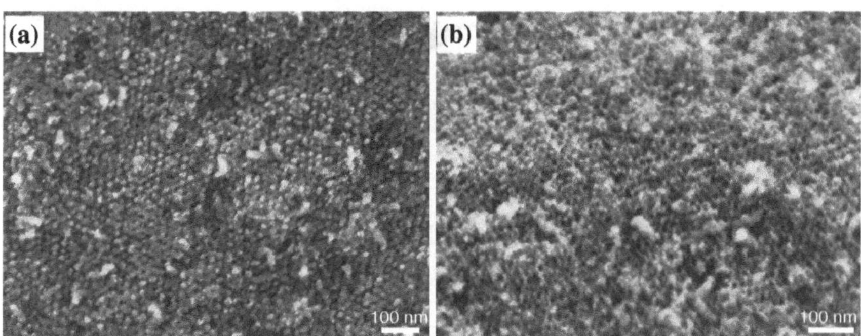

Fig. 6.6 **a** SEM image of the nickel gyroid coated with amorphous carbon by CVD. **b** SEM image of the hollow carbon gyroid after nickel etching

6.4 Inverse Gyroid

The fabrication of a structure similar to the hollow gyroid was attempted with the aim to reproduce the majority phase of the gyroid diblock copolymer and study the optical properties of such structure. Similarly to the hollow gyroid, this geometry divides the space into two separate parts and has two interfaces. Unfortunately the majority phase of the diblock copolymer is composed of polyfluorostyrene and can not be selectively etched, impeding its replication via the standard methodology. Nonetheless, its replication was attempted starting from the nickel double gyroid with a process shown in Fig. 6.7.

Gold was slowly electrodeposited onto the nickel strut network until it completely filled the void network, followed by overgrowth for several microns above the gyroid film. The overgrowth was removed by sonication in a similar fashion as described above. The nickel part was then selectively etched by $FeCl_3$ giving rise to the gold inverse structure of the double gyroid.

It is important to note that the sonication process cleaved the gold at the film-overgrowht boundary. This is presumably a consequence of the small grains and the high number of intergrain dislocations of the gold produced by electrodposition. Such a cleavage did not affect the nickel gyroid, which therefore acted as protective medium for the gold within the gyroid structure. It is also important to remark that several structural imperfections can be expected from this fabrication process, deriving from diffusion problems during electrodeposition preventing the complete filling of the gyroid void structure. The SEM image of the resulting structure is displayed in Fig. 6.8 together with the optical characterization.

Despite the presence of two interface in the inverse hollow gyroid, the transmission is significantly lower. This is a consequence of the high filling fraction of this structure of $f \approx 62\,\%$, the inverse of the double gyroid with $f = 38\,\%$. This value is however still higher than the transmission measured for the normal gyroid with $f = 60\,\%$, shown in Fig. 5.5a of the previous chapter.

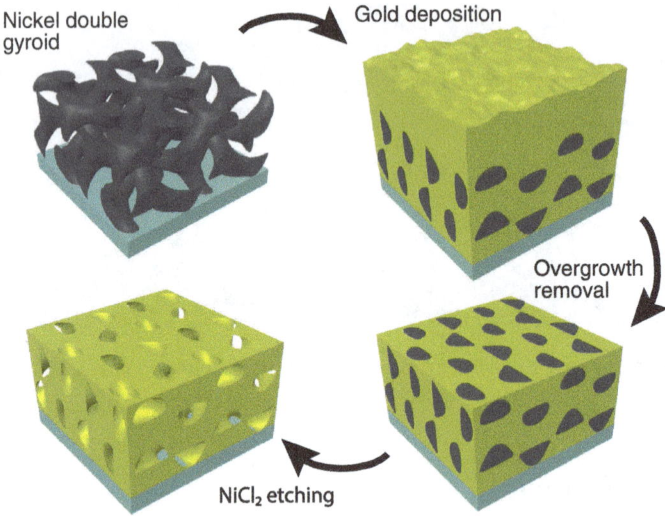

Fig. 6.7 Schematic representation of the inverse gyroid fabrication. Gold was electrodeposited onto the nickel double gyroid until complete filling followed by overgrowth of several microns above the film. Subsequently, the top overgrowth layer was removed by sonication. The nickel was then removed by $FeCl_3$ etching

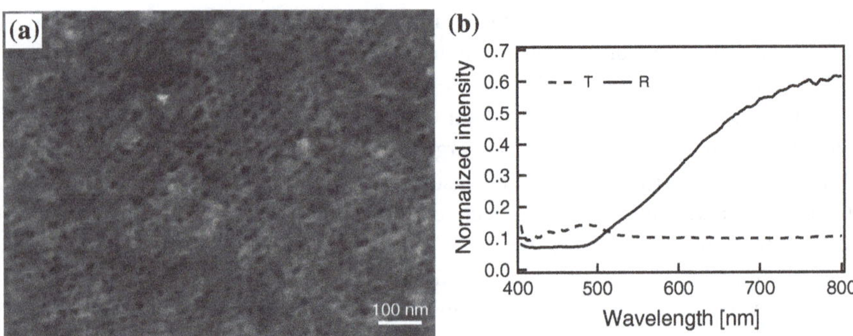

Fig. 6.8 **a** SEM image of the inverse gyroid structure. **b** Reflection and transmission spectra of the inverse gyroid

6.5 Conclusions

The complexity of the gyroid geometry was further increased by adding an internal interface through the manufacture of a hollow gyroid, producing a significant increase of reflection and transmission.

The hollow gyroid demonstrates that block copolymers can be used not only as a template to replicate their equilibrium phases but also as starting point to create more complex geometries. Moreover, the combination of different metals in the continuous structure opens new possibilities and gives an additional degree of freedom in the metamaterial design.

Further studies on the magnetic properties of the nickel-gold composite gyroid are under investigation and may soon lead to another interesting application of the gyroid metamaterial systems.

The fabrication of a carbon structure with gyroid geometry was successfully demonstrated and its application in batteries devices will be further investigated in our group. However, the long and relative high cost of fabrication may limit the practical applications of such a gyroid based batteries.

References

1. Goldstein J, Newbury DE, Joy DC, Lyman CE, Echlin P, Lifshin E, Sawyer L, Michael JR (2003) Scanning electron microscopy and X-ray microanalysis. Springer, New York
2. González-Díaz JB, García-Martín A, Armelles G, Navas D, Vázquez M, Nielsch K, Wehrspohn RB, Gösele U (2007) Enhanced magneto-optics and size effects in ferromagnetic nanowire arrays. Adv Mater 19(18):2643–2647
3. Valev VK, Silhanek AV, Gillijns W, Jeyaram Y, Paddubrouskaya H, Volodin A, Biris CG, Panoiu NC, De Clercq B, Ameloot M et al (2010) Plasmons reveal the direction of magnetization in nickel nanostructures. ACS Nano 5(1):91–96
4. Crozier KB, Sundaramurthy A, Kino GS, Quate CF (2003) Optical antennas: resonators for local field enhancement. J Appl Phys 94(7):4632–4642
5. Levin CS, Hofmann C, Ali TA, Kelly AT, Morosan E, Nordlander P, Whitmire KH, Halas NJ (2009) Magnetic-plasmonic core-shell nanoparticles. ACS Nano 3(6):1379–1388
6. Temnov VV, Armelles G, Woggon U, Guzatov D, Cebollada A, Garcia-Martin A, Garcia-Martin J-M, Thomay T, Leitenstorfer A, Bratschitsch R (2010) Active magneto-plasmonics in hybrid metal-ferromagnet structures. Nat Photonics 4(2):107–111
7. Zhang Q, Jiang T, Feng Y (2011) Slow-light propagation in a cylindrical dielectric waveguide with metamaterial cladding. J Phys D Appl Phys 44(47):475103
8. Kidambi PR, Ducati C, Dlubak B, Gardiner D, Weatherup RS, Martin M-B, Seneor P, Coles H, Hofmann S (2012) The parameter space of graphene chemical vapor deposition on polycrystalline Cu. J Phys Chem C 116(42):22492–22501
9. Endo M, Kim C, Nishimura K, Fujino T, Miyashita K (2000) Recent development of carbon materials for Li ion batteries. Carbon 38(2):183–197
10. Endo M, Kim YA, Hayashi T, Nishimura K, Matusita T, Miyashita K, Dresselhaus MS (2001) Vapor-grown carbon fibers (vgcfs) basic properties and their battery applications. Carbon 39(9):1287–1297

Chapter 7
Flexible and Stretchable Gyroid Metamaterials

7.1 Introduction

For versatile and practical applications, metamaterials are required to be robust and readily integrated to other systems, adapting their conformation to various geometries. For this reason stretchable and flexible metamaterials can be advantageous in many applications. Flexible metamaterials have been demonstrated at terahertz [1] and visible [2] frequencies with bidimensional structures.

However, state-of-art tridimensional optical metamaterials are still confined on rigid flat substrates and their nanostructure is easily exposed to structural damage.

In this chapter I will present the fabrication of stretchable and flexible gyroid metamaterials. The change of optical properties upon axial and plane strain deformations will also be discussed.

7.2 Stretchable Metamaterial

Stretchable metamaterials were fabricated by infiltrating an elastomer into the gold gyroid network and etching the sacrificial layer on which they were grown as depicted in Fig. 7.1. The glass substrate was initially coated with a 10 μm thick layer of nickel and functionalized with silane as described in Chap. 3 for ITO substrates. Gyroid block copolymers were then produced by thermal annealing and replicated into gold using the standard procedure.

A PS-*b*-PI-*b*-PS elastomer was dissolved in anisole (10 wt%, $M_{\mathrm{w}} = 8\,\mathrm{kg{\cdot}mol^{-1}}$) and infiltrated into the gyroid network by drop casting. The solvent was slowly evaporated at room temperature, leaving the gyroid metamaterial embedded in a thick transparent rubber matrix.

The infiltration process was strongly affected by the molecular weight of the elastomer. Long polymer chains, such as commercial polydimethylsiloxane (PDMS)

© Springer International Publishing Switzerland 2015

S. Salvatore, *Optical Metamaterials by Block Copolymer Self-Assembly*, Springer Theses,
DOI 10.1007/978-3-319-05332-5_7

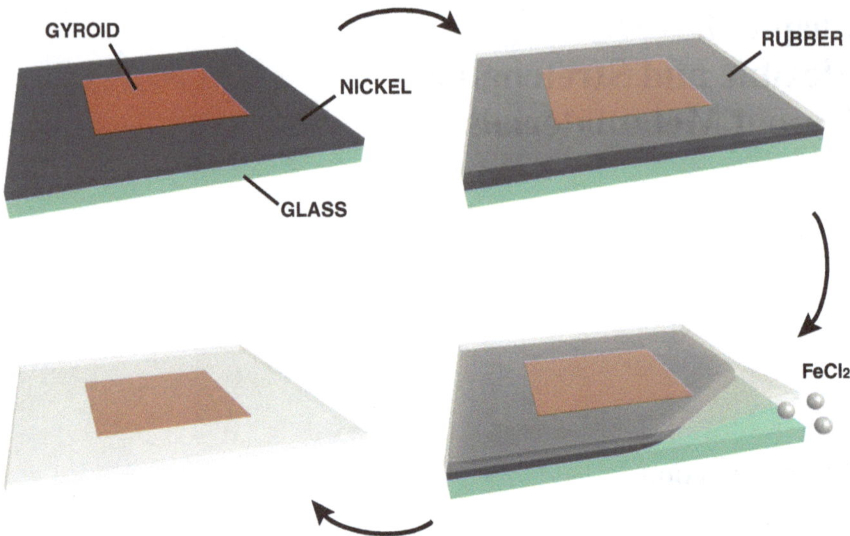

Fig. 7.1 Schematic of stretchable metamaterial fabrication. A gold gyroid was fabricated on a nickel-glass substrate and an elastomer was infiltrated into the gyroid network by drop casting. Finally, the nickel layer was etched way by iron chloride, leaving the gyroid metamaterial embedded in a transparent rubber matrix

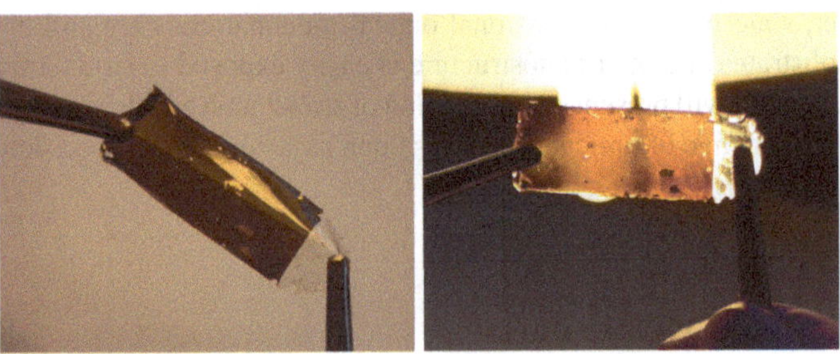

Fig. 7.2 Photos of the stretchable metamaterial in reflection and transmission. The width of the sample is approximately 4 cm

could not diffuse into the gyroid network due to the fact that the pore size of the gyroid network was smaller than the elastomer chain lengths.

The etching of the nickel layer allowed to obtain an optical metamaterial that was fully supported and embedded in an elastic and transparent matrix (Fig. 7.2).

To investigate the optical response of the elongated gyroid, the rubber film was extended using a mechanical stretcher, enabling the strain in one axial direction (and the compression in the orthogonal directions). As shown in Fig. 7.3, increasing stretching caused the reflection decrease at long wavelengths and the larger transmission band was slightly reduced.

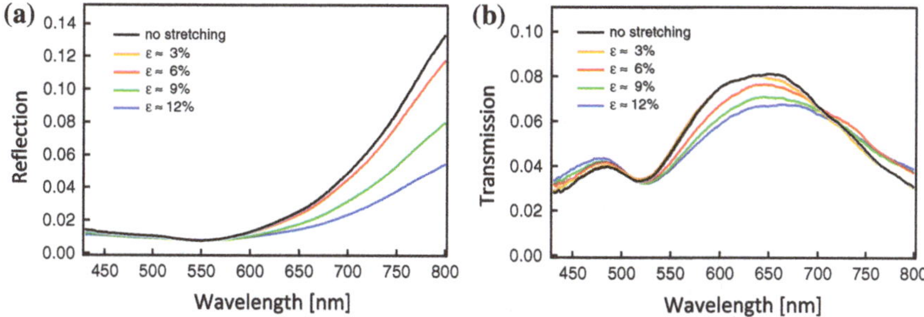

Fig. 7.3 Reflection and transmission spectra for unpolarized incident light on stretchable gyroid metamaterial for increasing strains

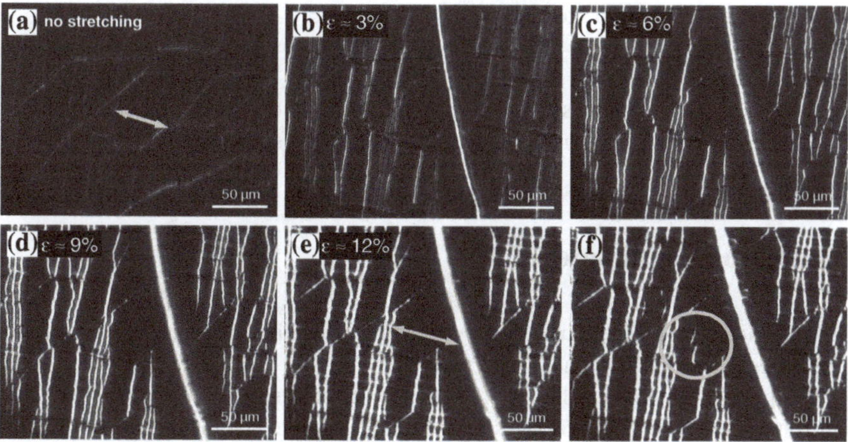

Fig. 7.4 Tranmsission images of a gyroid structure under axial stretching. **a** Initial state with no stretching, some discontinuities in the gyroid film were produced by residual rubber stress. **b–e** Stretching of the gyroid film and consequent elongation of the discontinuous blocks. **f** Cracks formation at a high strain

The gyroid film embedded in the rubber matrix, however, exhibited some discontinuities and fractures as a result of residual stresses in the elastomer. These were clearly visible in the transmission images shown in Fig. 7.4a. The dimensions of these discontinuous blocks (measured as distance from opposite edges) were monitored during film stretching and were effectively elongated in the strain direction (Fig. 7.4b–e) and compressed in the orthogonal direction, confirming the deformation of the gyroid structure (Fig. 7.4).

Large elongations induced additional fractures in the gyroid film, as shown in Fig. 7.4f that limited the study of the optical response at large deformations.

However, in this analysis, the optical measurements were averaged over different small gyroid domains. These have random in plane orientations, as discussed in the Chap. 3, and were therefore deformed along different axis during the film stretching.

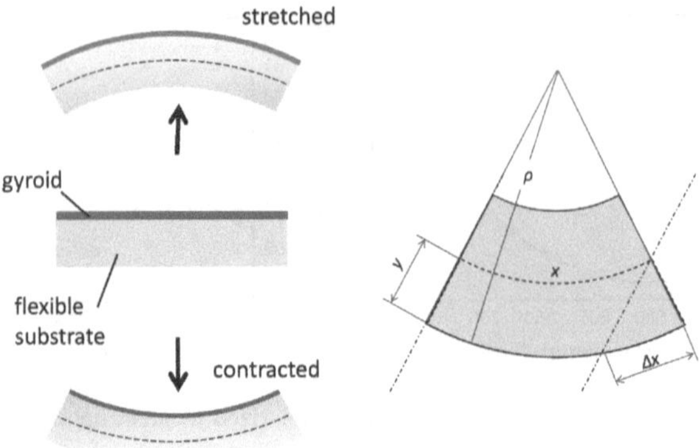

Fig. 7.5 Schematic of gyroid stretching by substrate bending

The study of the optical response of elongated single domains was conducted on gyroid films fabricated by drying-annealing on flexible substrates.

7.3 Flexible Metamaterial

Gold gyroids were fabricated by drying annealing on a 500 μm thick ethylene tetrafluoroethylene (ETFE) foil coated with nickel.

The gyroid metamaterial was elongated or compressed by bending the ETFE substrate as shown in Fig. 7.5. The strain could be extracted by the radius of curvature using the relation (see Appendix A.2)

$$\epsilon = \frac{\Delta x}{x} = y/\rho, \tag{7.1}$$

where ρ represents the radius of curvature, x is the neutral axis, and y is the distance from the neutral axis. In the elastic range, the tensile and compressive stress-strain curves of ETFE are identical and the neutral axis can therefore be approximated to be at one-half foil thickness.

The reflection spectra were measured on single gyroid domains for unpolarized and linearly polarized incident light. The domains were chosen with the in plane [110] orientation parallel and orthogonal to the elongation axis. As shown in Fig. 7.6, only minor variations were observed in the optical response for deformations of $\epsilon = 9\,\%$. The linear dichroism did not present significant variations in any of the elongated gyroid structures.

For deformations larger than $\epsilon = 9\,\%$, the gyroid film was fractured in the direction orthogonal to the strain axis. I note that these fractures occurred for smaller elongations than the gyroid stretched in the rubber matrix system. The reason may arise from the different strain states. In the rubber system the gyroid was elongated

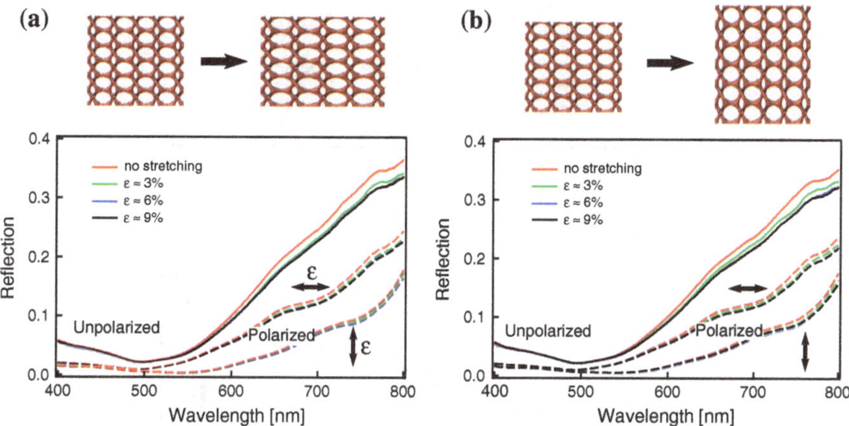

Fig. 7.6 Reflection spectra under plane strain for unpolarized and linearly polarized incident light. **a** Gyroid elongated along the in-plane [110] orientation. **b** Gyroid elongated along the in-plane [100] direction

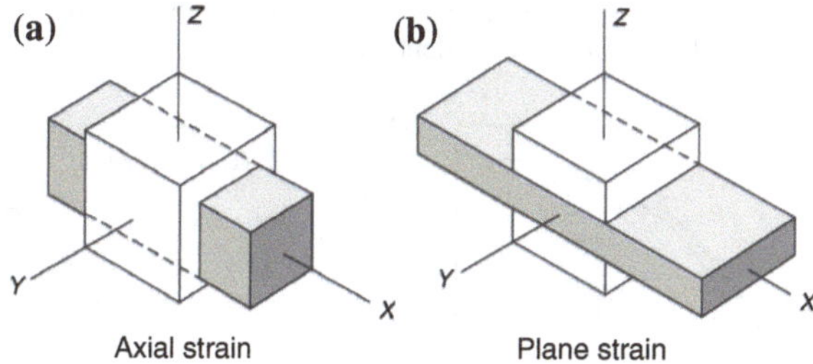

Fig. 7.7 Schematic of gyroid elongations. **a** Axial strain is induced in the rubber system stretching; **b** plane strain is a result of substrate bending

by axial strain (Fig. 7.7a), with consequent compression in the z and y directions, while on the curved substrate, the gyroid was elongated by plane strain (Fig. 7.7b), with no variations in the y direction and a larger compression in the z direction. In structural failure analysis the latter represents a more critical strain state due to the higher compression in the orthogonal direction [3].

Interestingly, larger optical variations were found in compressed gyroid as shown in Fig. 7.8. Reflection of polarized and unpolarized incident light resulted in a decrease of intensity with compression. However, the linear dichroism did again not show a significant change with deformation.

Simulation spectra from FDTD calculations were computed by Sang Soon Oh[1] for axial deformations and are presented in Fig. 7.9. These calculations suggested a significant variation of the optical response only for large deformations, which were

[1] Ortwin Hess' group, Department of Physics, Imperial College London.

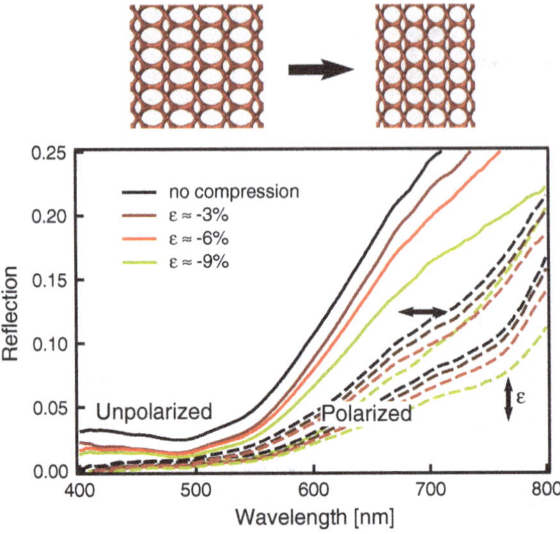

Fig. 7.8 Reflection spectra under plane strain for unpolarized and linearly polarized incident light. Gyroid compressed along the in-plane [110] orientation

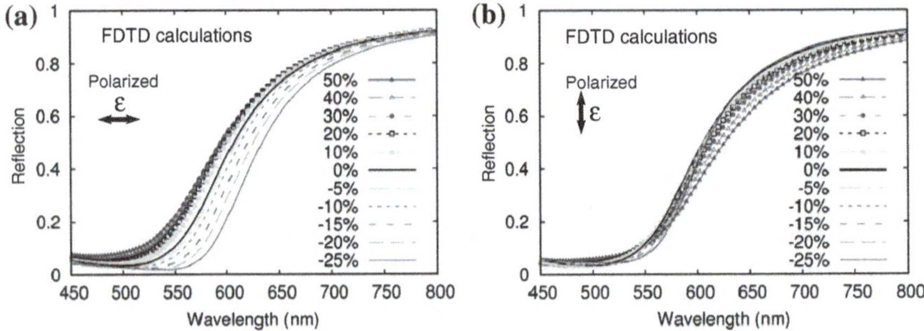

Fig. 7.9 FDTD calculations of a stretched and compressed gyroid structure under plane strain along the in-plane [110] orientation. Courtesy of Sang Soon Oh

experimentally unattainable because of fracture. However, the optical variations were predicted to be larger in the compressed gyroid, as shown experimentally.

7.4 Conclusions

The fabrication of stretchable and flexible metamaterials was demonstrated. This represents an important achievement for practical applications. This fabrication method can be easily implemented in other three-dimensional metamaterials.

However, the strain was limited to few percents before inducing structural fractures. The morphology of the elongated gyroid structure will also need to be studied by structural analysis (e.g. X-ray scattering) to investigate how the gyroid unit cell is modified under strain.

References

1. Lee S, Kim S, Kim T-T, Kim Y, Choi M, Lee SH, Kim J-Y, Min B (2012) Reversibly stretchable and tunable terahertz metamaterials with wrinkled layouts. Adv Mater 24(26):3491–3497
2. Di Falco A, Martin P, Thomas FK (2010) Flexible metamaterials at visible wavelengths. New J Phys 12(11):113006
3. Kazimi SMA (2001) Solid mechanics. Tata McGraw-Hill Education, New Delhi

Chapter 8
Metamaterial Sensors

8.1 Introduction

Gas sensing is an important process in a wide range of applications. In petrochemical industries sensors are used to ensure safety and monitor products [1]. Other applications include atmospheric science and breath diagnostics [2, 3]. The quantitative detections of gases is dominated by laboratory analytical equipment such as gas chromatographs, but these measurements preclude real time data acquisition. Semiconductor sensors can also be highly sensitive, but, they are specific to individual gases and have a limited lifetime.

Metamaterial have been proposed to improve the sensor performances by localized field enhancement [4]. For example, resonant modes of a 2D subwavelength resonator have been used for biosensing [5], and split ring resonator have been applied to monitor materials that respond resonantly at THz frequencies as explosives and DNA [6].

In this chapter I present the application of optical metamaterials as vapour sensors. This analysis offers a fast optical response at specific vapour concentrations and can be made in-situ without disturbing the sample environment.

It represents the first application of metamaterials for vapour sensing with immediate response at optical wavelengths, exploiting the high sensitivity of the metamaterial optical properties to the surrounding environment. This sensing system may fill an important gap between low cost sensors with inferior performance and highly sensitive laboratory equipment.

8.2 Capillary Condensation

The gyroid metamaterial sensor is based on the capillary condensation of liquid vapours into gyroid structures and on the resulting variation of their optical response. In capillary condensation, the vapour phase is adsorbed into a porous medium to the

© Springer International Publishing Switzerland 2015

S. Salvatore, *Optical Metamaterials by Block Copolymer Self-Assembly*, Springer Theses,
DOI 10.1007/978-3-319-05332-5_8

point at which the pore space are filled with condensed liquid from the vapour. The condensation occurs well below the saturation vapour pressure of the liquid. This is due to an increased number of van der Waals interactions between vapour phase molecules inside the confined space of the capillary [7].

Capillary condensation is an important factor in porous structures. It can be used to determine pore size distribution and surface area [8] and can be described by the Kelvin equation:

$$\ln \frac{P_v}{P_{sat}} = -\frac{2H\gamma V_l}{RT} \tag{8.1}$$

where P_v is the equilibrium vapour pressure, P_{sat} the saturation vapour pressure, H the mean pore curvature, γ the liquid/vapour surface tension, V_l the liquid molar volume, R the ideal gas constant, and T the temperature. Different pore geometries result in different types of curvature. The equations for the mean pore curvature H are available at numerous sources [9]. For a cylindrical capillary $H = 1/2r$, with r the radius of the cylinder.

The key aspect of gyroid structures in capillary condensation is the high monodispersity of their nanometric pores given by the ordered gyroid morphology. The capillary condensation can therefore be induced at very specific vapour pressures (i.e. vapour concentrations). Moreover, the gyroid unit cells and pore sizes can be easily varied with the block copolymer molecular weight.

The liquid condensation in gyroid metamaterials produces a shift in reflection towards longer wavelengths. As discussed in Chap. 5, this shift has a linear dependance to the liquid refractive index. This optical change enables the detection of vapour concentrations above a specific limit, set by the gyroid unit cell size.

8.3 Solvent Vapour Sensing

Gold gyroids with 35 and 50 nm unit cell sizes were first tested as sensors for water vapour.

The gold gyroids were loaded into a small chamber with a transparent window that allowed the optical characterization of the gyroid exposed to a solvent vapour (Fig. 8.1). The concentration of the solvent vapour was controlled by using two streams of nitrogen gas, one of which was passed through a washbottle filled with water, inducing the saturation of this stream before the two streams were mixed and introduced into the gyroid chamber. The flow rates of both streams were controlled by two electronic mass flow controllers, so that the precise adjustment of the relative flow rates controlled the relative solvent vapour pressure P_v/P_{sat} from 0 to 1, corresponding to water vapour saturation from 0 to 100 %.

The vapour pressure was increased in 1 % increments at intervals of 5 min to allow equilibration in the chamber. At $P_v/P_{sat} = 80\,\%$ the water vapour condensed in the smaller gyroid producing a red shift in reflection and change in coloration (Fig. 8.2),

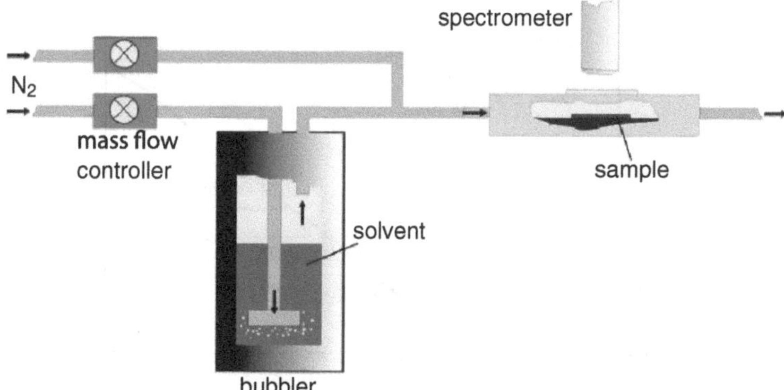

Fig. 8.1 Schematic of the solvent vapour chamber setup. Courtesy of Sven Hüttner

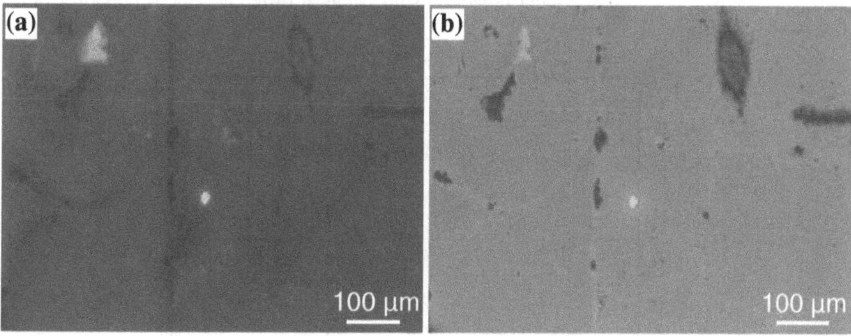

Fig. 8.2 Optical microscope images of the 35 nm unit cell gyroid before (**a**) and after (**b**) water vapour condensation

whereas the bigger gyroid were triggered at $P_v/P_{sat} = 85\,\%$. The corresponding reflection spectra are shown in Fig. 8.3.

A small hysteresis was found when decreasing the vapour pressure with water desorption occurring at vapour pressure 2 % lower than for adsorption. This deviation is characteristic for non-uniform pore geometries [10] where the adsorption takes place when the equilibrium pressure satisfies the larger pore section and the desorption is controlled by the smaller radius. It is reasonable to assume that the the variation of pore cross-section across the gyroid unit cell lies at the origin of the small experimental hysteresis.

The sensing response time could not be directly measured because of the dimension of the chamber and the small flow rates. However, the desorption process was immediate (less than 1 s) upon the opening of the chamber, proving the very fast response of this sensing system.

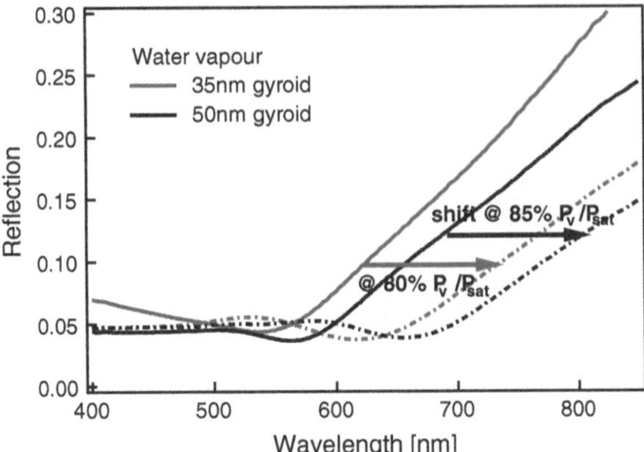

Fig. 8.3 Reflection spectra of the gyroid metamaterial with increasing water vapour pressure. The capillary condensation and consequent reflection shift occur at different vapour pressures in the smaller and bigger gyroid

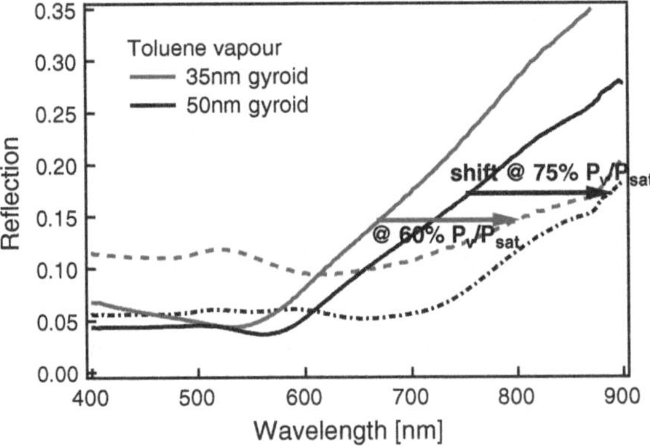

Fig. 8.4 Reflection spectra of the gyroid metamaterial with increasing toluene vapour pressure. The capillary condensation and consequent reflection shift occur at different vapour pressures in the smaller and bigger gyroid

Similar tests were performed with toluene vapour. In this case the smaller gyroid detected a vapour pressure of $P_v/P_{sat} = 60\,\%$ and the bigger gyroid of $75\,\%$ (Fig. 8.4).

These results can be compared with the values from Eq. 8.1. The surface tensions of water and toluene were 72.8 and 28.4 mN/m [11], respectively. The mean pore curvature was approximated as one-half the difference of gyroid unit cell and strut thickness. As shown in Table 8.1 the capillary condensation was expected to occur at higher vapour pressures. This can be addressed by modifying the radius of curvature approximation. The gyroid pores were approximated as cylinders, whereas Kelvin equation should be modified to take into account the gyroid pore geometry.

Table 8.1 Comparison of experimental and predicted values of vapour pressures sensed by gyroids with different unit cell sizes

		P_v/P_{sat}	
Vapour	Unit cell size (nm)	Experimental (%)	Kelvin eq. (%)
Water	35	80	86
	50	85	91
Toluene	35	60	71
	50	75	81

However, as found experimentally, lower surface tension decreases the vapour pressure of capillary condensation and increases its variation with pore size. In industrial applications a number of monitored vapours have low surface tensions, such as ammonia, benzene, formaldehyde, sulfur dioxide, supporting the appeal of this gyroid metamaterial sensor.

8.4 Conclusions

The application of an optical metamaterial as vapour sensor was successfully proven. Environmental parameters strongly affect the optical response of the optical metamaterial and it was here exploited to monitor the concentration of water and toluene vapours. This sensor system may be applied to a number of other solvents, tuning the sensing response by the initial block copolymer molecular weight.

References

1. Allen MG (1998) Diode laser absorption sensors for gas-dynamic and combustion flows. Measur Sci Tech 9(4):545
2. Laj P, Klausen J, Bilde M, Plaß-Duelmer C, Pappalardo G, Clerbaux C, Baltensperger U, Hjorth J, Simpson D, Reimann S et al (2009) Measuring atmospheric composition change. Atmos environ 43(33):5351–5414
3. Smith D, Španěl P (2007) The challenge of breath analysis for clinical diagnosis and therapeutic monitoring. Analyst 132(5):390–396
4. Jakšić Z, Vuković S, Matovic J, Tanasković D (2010) Negative refractive index metasurfaces for enhanced biosensing. Materials 4(1):1–36
5. He S, Jin Y, Ruan Z, Kuang J (2005) On subwavelength and open resonators involving metamaterials of negative refraction index. N J Phys 7(1):210
6. Fischer BM, Walther M, Jepsen PU (2002) Far-infrared vibrational modes of DNA components studied by terahertz time-domain spectroscopy. Phys Med Biol 47(21):3807
7. Schramm LL (1993) The language of colloid and interface science: a dictionary of terms. American Chemical Society, Washington

8. Kruk M, Jaroniec M, Sayari A (1997) Application of large pore MCM-41 molecular sieves to improve pore size analysis using nitrogen adsorption measurements. Langmuir 13(23):6267–6273
9. Miyahara M, Kanda H, Yoshioka T, Okazaki M (2000) Modeling capillary condensation in cylindrical nanopores: a molecular dynamics study. Langmuir 16(9):4293–4299
10. Hunter RJ, White LR, Chan DYC (1987) Foundations of colloid science. Clarendon Press Oxford, New York
11. Haynes WM, Lide DR, Bruno TJ (2012) CRC Handbook of chemistry and physics 2012–2013. CRC Press, Boca Raton

Chapter 9
Conclusions

The fabrication, characterization, developments and practical applications of optical metamaterials made by block copolymer self-assembly were successfully demonstrated.

The achievements obtained in the fabrication process were remarkable. The fabrication of nanometric features in complex three-dimensional metallic structures represent a major obstacle to create metamaterials active at visible frequencies. In the gyroid metamaterials 10 nm features were obtained with a high reproducibility and by self-assembly, with a precise control of the structural parameters. This is, to date, the three-dimensional metamaterial with the smallest feature size ever made.

Moreover, the good reproducibility achieved with the fabrication of gyroid metamaterial underlines the potential of the bottom-up approach by block copolymer self-assembly in producing optical metamaterials.

The complexity of the gyroid geometry was further increased by adding an internal interface through the manufacture of a hollow gyroid. The combination of different metals in the continuous hollow-core 'composite' metamaterial opens new possibilities and gives an additional degree of freedom in the metamaterial design.

Gyroid metamaterials have shown interesting optical properties such as reduced plasma frequency, enhanced optical transmission and optical tunability. The propagation of the plasmonic modes through the gold gyroid structure was largely demonstrated and it represents an hallmark of optical metamaterials. Other optical properties such as fast response of non linear properties of gold gyroid have also been investigated in collaboration with Petros Farah. These results were not discussed in this dissertation and have been submitted to Physical Review Letter.

Flexible and stretchable metamaterial were fabricated by infiltrating an elastomer around the gyroid network. Although only small strains could be applied before structure deformation, these systems represent an important step for future real application of optical metamaterials.

Finally, the gyroid metamaterial was used as vapour sensor, exploiting the regular pore size and the variation of the optical response with surrounding media. This represents the first application of metamaterials for vapour sensing with immediate response at optical wavelengths.

© Springer International Publishing Switzerland 2015

S. Salvatore, *Optical Metamaterials by Block Copolymer Self-Assembly*, Springer Theses,
DOI 10.1007/978-3-319-05332-5_9

A further work on the gyroid metamaterials could be focused on the investigation of the refractive index value. A direct measurements might be achieved on the gyroid films produced with the grating morphology discussed in Chap. 3. However, the control of the gyroid orientation in thin films is paramount to fully investigate the optical properties in different directions.

A preliminary study on the surface coupling was performed in a Kretschmann configuration with an hemisphere lens set-up illuminated via a confocal microscope and imaged in the back focal plane. Different coupling intensities were observed in different in-plane directions, demonstrating different propagation modes through different gyroid orientations. However, a quantitative study will be necessary and not trivial to lead to a full experimental optical characterization.

The gyroid morphology is often considered of low industrial interest in polymer science, due to the tiny portion occupied in the phase diagram that obligate a very controlled and expensive polymer synthesis process. Nonetheless, in the metamaterial field it is still a very appealing and low-cost approach compared to the conventional top-down processes such as e-bean lithography and, therefore, an ideal candidate for mass production of optical metamaterials.

An experimental proof of the negative refractive index, together with the optical characterization and the fabrication developments shown in this work, could strongly support the application of the gyroid structure as optical metamaterials and, maybe, make a step closer to Orwell's visionary novel.

Appendix

A.1

The gyroid is consisted of multiple helices [1], with a 2D graphic representation shown in Fig. A.1a. The continuous nature of the metallic helices give rise to a cut-off wavelength, the plasma wavelength. No wave can propagate in the gyroid for wavelengths larger than the plasma wavelength. Conversely, there are three propagating modes just below the plasma wavelength and using FDTD simulations [1] it was possible to calculate their field profiles (Fig. A.1b). These modes are highly confined within the smaller helices and very weakly interact with the larger helices. Hence, we can approximate the complex gyroid structure to a tri-helical metamaterial (THM) made from just the small helices. Figure A.1a shows the geometric parameters of the small gyroid helices, where R and b are the radius and the lattice constant, respectively, a is the periodicity of the gyroid, and r_w is the radius of the wire forming the small helices.

The analytical derivation of the tri-helical model is described in detail in Demetriadou et al. [2]. Here, I present the effective electromagnetic parameters of a THM structure that are dispersive and have Lorentzian and Drude-like forms. The inverse effective parameters are given by

$$\chi_{\mathrm{EE}}^{-1} = \left(\frac{b}{l}\right)\left(\frac{\omega^2}{\omega^2 - \omega_{\mathrm{p}}^2}\right), \tag{A.1}$$

$$\chi_{\mathrm{HH}}^{-1} = \left(\frac{b}{l}\right)\left(\frac{1 + L\,(a/\,(2\pi R))}{1 - \pi R^2/a^2 + La/\,(2\pi R)}\right)\left(\frac{\omega^2 - \omega_{0m}^2}{\omega^2 - \omega_p^2}\right), \tag{A.2}$$

$$\kappa_{\mathrm{EH}}^{-1} = c_0^2\left(\frac{b}{l}\right)\left(\frac{iG\omega_p^2\omega}{\omega^2 - \omega_p^2}\right) = \kappa_{\mathrm{HE}}^{-1}, \tag{A.3}$$

© Springer International Publishing Switzerland 2015

S. Salvatore, *Optical Metamaterials by Block Copolymer Self-Assembly*, Springer Theses,
DOI 10.1007/978-3-319-05332-5

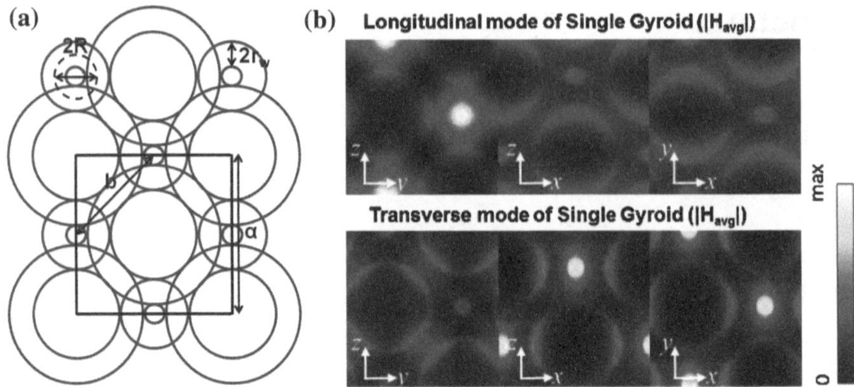

Fig. A.1 **a** A 2D-graphic representation of the gyroid topology, where R is the radius of the small (*red*) helices, r_w is the radius of the wires forming the gyroid, a is the periodicity of the gyroid and b the lattice constant of the small helices. **b** The field profiles for both the longitudinal (*top figures*) and transverse modes (*bottom figures*) for propagation along the x-axis

where χ_{EE}^{-1} and χ_{HH}^{-1} are the inverse electric permittivity and magnetic permeability respectively, $\kappa_{EH}^{-1} = \chi_{EH}^{-1}/\mu_0$, $\kappa_{EH}^{-1} = \chi_{EH}^{-1}/\epsilon_0$ are the inverse chirality terms, and $l = \sqrt{(2\pi R)^2 + a^2}$ is the length of the helix per unit cell. The non-inverse parameters are derived from

$$\begin{bmatrix} \epsilon_0\chi_{EE} & \chi_{EH} \\ \chi_{HE} & \mu_0\chi_{HH} \end{bmatrix} = \begin{bmatrix} \chi_{EE}^{-1}/\epsilon_0 & \chi_{EH}^{-1} \\ \chi_{HE}^{-1} & \chi_{HH}^{-1}/\mu_0 \end{bmatrix}^{-1} = \frac{1}{\det}\begin{bmatrix} \chi_{HH}^{-1}/\mu_0 & -\chi_{EH}^{-1} \\ -\chi_{HE}^{-1} & \chi_{EE}^{-1}/\epsilon_0 \end{bmatrix}, \tag{A.4}$$

where $\det = c_0^2\chi_{EE}^{-1} + \left(\kappa_{EH}^{-1}\right)^2/c_0^2$.

The plasma frequency (ω_p) and the magnetic resonance frequency (ω_{0m}) are given by

$$\omega_p = c_0\sqrt{\frac{a^2}{\pi R^2 b^2\left(1 - \pi R^2/b^2 + La/(2\pi R)\right)^2}};\quad \omega_{0m} = c_0\sqrt{\frac{a^2}{\pi R^2 b^2\left(1 + La/(2\pi R)\right)^2}}, \tag{A.5}$$

where $G = \pi R^2/a$ is a constant and $L = (l/b)\,ln\sqrt{b^3/\left(\pi r_w^2 l\right)}$ the self-inductance of the wires per unit cell. Therefore, the electric permittivity of the gyroid is given by $\chi_{EE} = \frac{\chi_{HH}^{-1}}{\chi_{EE}^{-1}\chi_{HH}^{-1} + \left(\kappa_{EH}^{-1}\right)^2/c_0^4}$, and since $\left(\kappa_{EH}^{-1}\right)^2/c_0^4 \to 0$,

$$\chi_{EE} \approx \frac{1}{\chi_{EE}^{-1}} = \left(\frac{l}{b}\right)\left(\frac{\omega^2 - \omega_p^2}{\omega^2}\right) = \left(\frac{l}{b}\right)\left(1 - \frac{\omega_p^2}{\omega^2}\right). \tag{A.6}$$

We have however so far neglected the field penetration into the metallic wires and the resistivity losses of the metal. Since all the gyroid structures studied in this paper are made from nano-scale wires, their metallic parts are completely penetrated by the fields. These fields induce currents not only on the surface of the metallic wires, but also inside the metal as described in ref. [3]. This phenomenon significantly red-shifts the optical response of the helical wires and can be accounted for in the above analytical model using

$$\omega = \frac{\omega'}{\sqrt{\frac{(\omega' r_{\mathrm{w}} \tilde{z})^2}{c_0^2 + \epsilon_n (\omega' r_{\mathrm{w}} \tilde{z})^2} - \frac{4 r_{\mathrm{w}} \omega'}{2\pi c_0}}}, \tag{A.7}$$

where ϵ_n is the host medium of refractive index n and $\tilde{z}$ is a function of ϵ_n and ϵ_{m}, the permittivity of metal [3]. In order to find the effective electric permittivity of the gyroid made from nano-scale metallic components, we replace all ω of Eq. (A.6) with Eq. (A.7). At the plasma frequency $\left(\omega'_{\mathrm{p}}\right)$ or wavelength $\left(\lambda'_{\mathrm{p}}\right)$, the effective electric permittivity is $\chi_{EE} = 0$. Applying these conditions to Eq. (A.6), the plasma frequency $\left(\omega'_{\mathrm{p}}\right)$ is

$$\frac{\omega'_{\mathrm{p}}}{\sqrt{\frac{\left(\omega'_{\mathrm{p}} r_{\mathrm{w}} \tilde{z}\right)^2}{c_0^2 + \epsilon_n \left(\omega'_{\mathrm{p}} r_{\mathrm{w}} \tilde{z}\right)^2} - \frac{4 r_{\mathrm{w}} \omega'_{\mathrm{p}}}{2\pi c_0}}} = \omega_{\mathrm{p}} \quad \Rightarrow \quad \lambda'_{\mathrm{p}} \sqrt{\frac{\left(\omega'_{\mathrm{p}} r_{\mathrm{w}} \tilde{z}\right)^2}{c_0^2 + \epsilon_n \left(\omega'_{\mathrm{p}} r_{\mathrm{w}} \tilde{z}\right)^2}} = \lambda_{\mathrm{p}} + 4 r_{\mathrm{w}}, \tag{A.8}$$

where $\lambda_{\mathrm{p}} = 2\pi c_0/\omega_{\mathrm{p}} = 2\pi \Big/ \sqrt{\frac{a}{\pi R^2 b^2 (1 - n R^2/b^2 + L a (2\pi R))}}$. The square root of Eq. (A.8) can be approximated for gold to

$$\sqrt{\frac{\left(\omega'_{\mathrm{p}} r_{\mathrm{w}} \tilde{z}\right)^2}{\left(c_0^2 + \epsilon_n \left(\omega'_{\mathrm{p}} r_{\mathrm{w}} \tilde{z}\right)^2\right)}} \approx \frac{\left(\frac{\tilde{z} r_{\mathrm{w}}}{\sqrt{2\epsilon_n}}\right)}{\sqrt{\frac{r_{\mathrm{w}}^2 \tilde{z}^2}{2} - \frac{\lambda_{\mathrm{p}}'^2}{4\pi^2 \epsilon_m \left(\lambda'_{\mathrm{p}}\right)}}}, \tag{A.9}$$

where the full expression for $\tilde{z}$ can be found in ref. [3]. For the gold gyroids discussed in this letter, the permittivity is modelled as

$$\epsilon_{\mathrm{Au}}(\lambda) = \epsilon_\infty - \frac{\lambda^2}{\lambda_{\mathrm{Au}}^2 \left(1 + i g_{\mathrm{Au}} \lambda / 2\pi c_0\right)} + \Sigma_{n=1}^2 \frac{\sigma_n \lambda^2}{\lambda^2 - i g_n \lambda \left(\lambda_{\mathrm{p}}^2\right)_n - \left(\lambda_{\mathrm{p}}^2\right)_n}, \tag{A.10}$$

where $\epsilon_\infty = 4.9752$, $\lambda_{\mathrm{Au}} = 128.53\,\mathrm{nm}$, $g_{\mathrm{Au}} = 32.63\,\mu\mathrm{m}$, $\sigma_1 = 1.76$ $\sigma_2 = 0.952$, $\left(\lambda_{\mathrm{p}}\right)_1 = 338.17\,\mathrm{nm}$, $\left(\lambda_{\mathrm{p}}\right)_2 = 437.01\,\mathrm{nm}$, $g_1 = 946.61\,\mathrm{nm}$, $g_2 = 1.6833\,\mu\mathrm{m}$.

For gold structures with ϵ_{Au}, $\tilde{z}$ can be approximated to $\tilde{z} \approx -C\left(\frac{a_0}{2+a_0}\right)\left(\frac{2+\sqrt{2}}{2}\right) \sim 1.0788$, where $a_0 = \Gamma - \ln 2$, $\Gamma = 0.577$ is the Euler constant and $C \sim 10.25$ is a constant. Also, for a gyroid of periodicity a and filling fraction f, the small helices have a radius $R = a\left(2-\sqrt{2}\right)/4$, a lattice constant $b = a/\sqrt{2}$ and a radius of the metallic wires of $r_{\mathrm{w}} = \sqrt[4]{2}a/\sqrt{\pi\left(\sqrt{2+\pi^2}+\sqrt{2+\left(3-2\sqrt{2}\right)\pi^2}\right)} \approx 0.2896a\sqrt{f}$. Hence, Eq. (A.8) becomes

$$\lambda_{\mathrm{p}} = \lambda_{\mathrm{Au}} \approx c_1 a\left(\frac{1}{\epsilon_{\mathrm{Au}} f c_2} + \frac{c_3}{\epsilon_n\left(c_4\sqrt{f} - \sqrt{c_5 + c_6 \ln\left(c_7/\sqrt{f}\right)}\right)^2}\right)^{-\frac{1}{2}}, \quad \text{(A.11)}$$

where a and f are the periodicity and filling fraction of the gyroid, respectively, and c_{1-7} are the geometrical coefficients given by

$$\begin{aligned}
c_1 &= 2^{\frac{3}{4}}\sqrt{\frac{\pi}{\sqrt{2+\pi^2}+\sqrt{2+(3-2\sqrt{2})\pi^2}}},\\
c_2 &= \tilde{z} \approx -C\left(\frac{a_0}{2+a_0}\right)\left(\frac{2+\sqrt{2}}{2}\right),\\
c_3 &= 128\pi^2,\\
c_4 &= -32,\\
c_5 &= -\sqrt{2}\pi^4(-2+\sqrt{2})\left(8-4\sqrt{2}+(-10+7\sqrt{2})\pi\right)\left(\sqrt{2+\pi^2}+\sqrt{2+(3-2\sqrt{2})\pi^2}\right),\\
c_6 &= -8\pi^3(-2+\sqrt{2})\sqrt{2(2+(3-2\sqrt{2})\pi^2)}\left(\sqrt{2+\pi^2}+\sqrt{2+(3-2\sqrt{2})\pi^2}\right),\\
c_7 &= \frac{\sqrt{1+\sqrt{\frac{2+\pi^2}{2+(3-2\sqrt{2})\pi^2}}}}{2^{\frac{3}{4}}}.
\end{aligned} \quad \text{(A.12)}$$

Equation (A.11) shows that the plasma wavelength depends strongly on the dispersion of the metal permittivity, the surrounding dielectric medium, the periodicity, and the filling fraction of the gyroid.

A.2

An incremental element of a beam is shown in Fig. A.2 for undeformed and deformed conditions. By definition the normal strain along Δs is determined as:

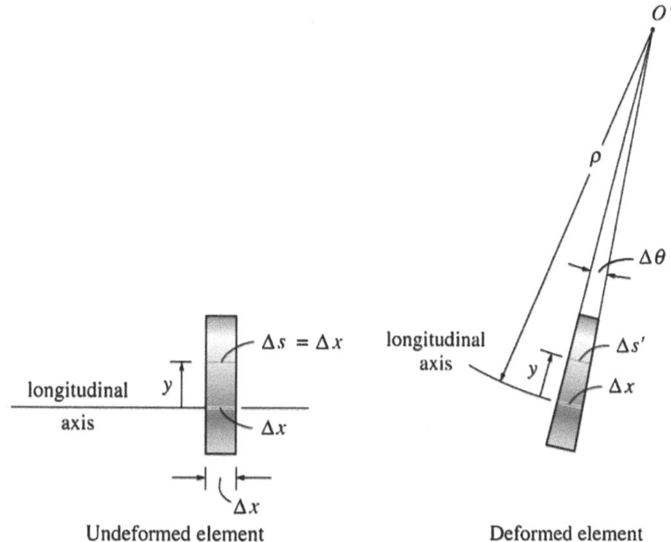

Fig. A.2 Undeformed and deformed elements in a beam. The longitudinal axis is set with respect to the neutral axis

$$\epsilon = \lim_{\Delta s \to 0} \frac{\Delta s' - \Delta s}{\Delta s}. \tag{A.13}$$

The strain can be represented in terms of distance y from the neutral axis and radius of curvature ρ. Since $\Delta\theta$ defines the angle between the cross sectional sides of the incremental elements, $\Delta s = \Delta x = \rho\Delta\theta$. The deformed length of Δs becomes $\Delta s' = (\rho - y)\Delta\theta$. Introducing these relations into Eq. (A.13) gives:

$$\epsilon = \lim_{\Delta s \to 0} \frac{(\rho - y)\Delta\theta - \rho\Delta\theta}{\rho\Delta\theta} = \frac{-y}{\rho}. \tag{A.14}$$

References

1. Soon Oh S, Demetriadou A, Wuestner S, Hess O (2012) On the origin of chirality in nanoplasmonic gyroid metamaterials. Adv Mater 25(4):612–617. doi:10.1002/adma.201202788
2. Demetriadou A, Soon Oh S, Wuestner S, Hess O (2012) A tri-helical model for gyroid metamaterials. New J Phys 14(083032):724–726. doi:10.1088/1367-2630/14/8/083032
3. Demetriadou A, Hess O (2013) Theory of optical nano-plasmonic metamaterials. Phys Rev B 87(16):161101-1–161101-5

Zeitfracht Medien GmbH
Ferdinand-Jühlke-Straße 7
99095 Erfurt, Deutschland
produktsicherheit@kolibri360.de